The ...ton
Review

...princetonreview.com

Know It All!
Grades 9–12 Math

by James Flynn

Random House, Inc.
New York

www.randomhouse.com/princetonreview

This workbook was written by The Princeton Review, one of the nation's leaders in test preparation. The Princeton Review helps millions of students every year prepare for standardized assessments of all kinds. The Princeton Review offers the best way to help students excel on standardized tests.

The Princeton Review is not affiliated with Princeton University or Educational Testing Service.

Princeton Review Publishing, L.L.C.
160 Varick Street, 12th Floor
New York, NY 10013

E-mail: textbook@review.com

Published in the United States by Random House, Inc., New York.

ISBN 0-375-76377-5

Editor: Michael Bagnulo
Development Editor: Scott Bridi
Production Editor: Evangelos Vasilakis
Director of Production: Iam Williams
Design Director: Tina McMaster
Art Director: Neil McMahon
Production Manager: Mike Rockwitz

Manufactured in the United States of America

9 8 7 6 5 4 3 2 1

First Edition

Contents

Introduction

Introduction

About This Book

What kind of person is a *know it all*? Someone who craves information and wants to learn new things. Someone who wants to be amazed by what they learn. Someone who is excited by the strange and unusual.

Know It All! is an adventure for your mind. *Know It All!* is full of weird, fascinating, unbelievable articles—all of which contain true stories and actual events!

In addition to feeding your brain with all sorts of interesting information, *Know It All!* will feed your brain with test-taking tips and standardized-test practice.

Know It All! contains chapters, review sections called "Brain Boosters," an answer key for the chapters and Brain Boosters, a practice test, and answers and explanations for the practice test.

Each **chapter**

- defines a skill or group of skills

- shows how to use this skill or group of skills to answer a sample question

- provides practice passages and questions similar to those you may see on standardized tests in school

Each **Brain Booster**

- reviews skills in the previous chapters

- includes interesting passages to read

- provides practice questions to answer

The **answer key** provides correct answers to the questions in the chapters and brain boosters.

The **practice test**

- gives you practice answering questions similar to those on standardized-achievement tests

- provides a bubble sheet that is similar to the type you'll see on standardized-achievement tests

The **answers and explanations** provide correct answers to the questions on the practice test and explanations for correct answers.

About Standardized-Achievement Tests

You know about them. You've probably taken them. But you might have a few questions about them. If you want to be a *know it all,* then it would be good for you to know about standardized-achievement tests.

The words *standardized* and *achievement* describe the word *tests. Standardized* means to compare something with a standard. Standardized tests often require you to answer questions about standards that have been decided by your school, district, or state. These standards list the skills you will learn in different subjects in different grades. *Achievement* means the quality of the work produced by a student. So *standardized-achievement tests* assess the quality of your work with certain skills.

To find out the nitty-gritty about any standardized-achievement test you may take, ask your teachers and parents about the tests. The following are some questions you might want to ask.

Who?	You!
What?	What kinds of questions will be on the test?
	What kinds of skills will be tested by the test?
Where?	Where will the test be given?
When?	When will I take the test?
	How much time will I have to complete the test?
Why?	Why am I taking the test?
	Do I have to pass the test to graduate?
How?	How should I prepare for the test?
	Do I need to bring anything to the test?

No test can assess your unique qualities as a *know it all.* The purpose of a standardized-achievement test is to show how well you can use the skills that you learned in school in a testing situation.

About The Princeton Review

The Princeton Review is one of the nation's leaders in test preparation. We prepare more than two million students every year with our courses, books, online services, and software programs. We help students around the country on many statewide and national standardized tests as well as college-entrance exams such as the SAT-I, SAT-II, and ACT. Our proven strategies and techniques reinforce the skills students learn in the classroom and help them apply these skills to standardized tests.

Things to Remember When Preparing for Tests

There are lots of things you can do to prepare for standardized-achievement tests. Here are a few examples.

- **Work hard in school all year.** Working hard in school all year is a great way to prepare for tests.

- **Read.** Read everything you can. Read your homework, your textbooks, the newspapers, magazines, novels, plays, poems, comics, and even the back of the cereal box. Reading a lot is a great way to prepare for tests.

- **Work on this book!** This book provides you with lots of practice for tests. You've probably heard the saying, "Practice makes perfect." Practice is a great way to prepare for tests.

- **Ask your teachers and parents questions about your schoolwork whenever you need to.** Your teachers and parents can help you with your schoolwork. Asking for help when you need it is a great way to prepare for tests and to become a *know it all*.

- **Ask your teachers and parents for information about the tests.** If you have questions about the tests, ask them! If you are informed about a test, you will be prepared for it.

- **Have a good dinner and a breakfast before you take a test.** Eating well will fuel your body with energy, and your brain thrives on energy. You want to take each test with all the engines in your brain running properly.

- **Get enough sleep before you take a test.** Being awake and alert while taking tests is very important. Your body and mind work best when you've had enough sleep. So get plenty of Zs on the nights before the tests!

- **Check your work.** When taking a test, you may finish it with extra time left. You could spend that extra time twiddling your thumbs or timing how long you can go without blinking. But if you use that extra time to check your work, you might spot some mistakes—and improve your score.

- **Stay focused.** You may find that your mind might wander away from the test once in a while. Don't worry—it happens. Just say to your brain, "Brain, it's great that you are so curious, imaginative, and energetic, but I need to focus on one thing right now—the test." Your brain will thank you later.

About the Icons in This Book

This book contains many different small pictures, called *icons*. The icons tell you about the topics in the articles in the book.

 Alternative Animals Read these passages to learn about animals you never knew existed and feats you never knew animals could accomplish. You'll learn about the biggest, smallest, oldest, fastest, and most interesting animals on the planet.

 Hip History Your mission is to storm some of the coolest castles in history with some extraordinary historic figures—some of whom aren't much older than you. These passages will help you complete that mission while learning some of the most interesting stories in history.

 For Your Amusement You want to play games? Read these passages to learn about cool games, toys, amusement parks, and festivals.

 Extreme Sports Read these passages to learn about outrageous contests, wacky personalities, and incredible feats in the world of sports.

 Grosser Than Gross How gross can you get? Read these passages if you want to learn about really gross things. Be warned: Some of the passages may be so gross that they're downright scary.

 Mad Science If you read these passages, you'll see science like you've never seen it before. You'll learn about all sorts of interesting scientific stuff.

Outer Space Oddities Do you ever wonder what goes on in the universe away from planet Earth? Satisfy your curiosity by reading these passages about astronomical outer space oddities.

 Explorers and Adventurers Did you ever want to take a journey to learn more about a place? Well, you'll get the opportunity to do that if you read these passages about explorers and adventurers.

 The Entertainment Center Do you enjoy listening to music or watching television and movies? Well, here's your chance to read about them!

 Art-rageous Are you feeling a bit creative? Read these passages to get an unusual look into art that's all around you: books, drawings, paintings, and much more.

 Bizarre Human Feats People do some strange and incredible stuff. You can read about some of these incredible-but-true human feats in these passages.

 Wild Cards You'll never know what you're going to get with these passages. It's a mixed bag. Anything goes!

The Chapters

Chapter 1: Question Types

Why are working teenagers paid less than the minimum wage?

What are the chances of being struck by lightning?

Why is the thirteenth floor of a building a "lucky" floor?

Math tests, whether they are national or statewide, usually require students to answer three different types of questions: multiple choice, short answer, and open response.

Multiple-Choice Questions

Multiple-choice questions give you a question and then offer several answer choices. The best thing about multiple-choice questions is that you do not have to come up with the correct answer on your own. It is printed right there on the paper in front of you! The correct answer is **always** one of the answer choices. Here's the **Know It All Approach** to answering multiple-choice questions.

Step 1   **Read the question carefully.** This should be the first step for answering **any** question. By reading the question carefully, you can figure out exactly what is being asked, and then consider what you have to do in order to find the correct answer. It may help you to notice (and perhaps circle or underline) key words such as **not, except, each, total, least, greatest, exact, approximate,** and **estimate.** You should also pay particular attention to numbers, variables, formulas, and equations that appear in the text of the question. Most likely, you will have to use them to figure out the correct answer.

Step 2   **Answer the question.** You're probably thinking, "No kidding!" However, this is an important step. If you need to perform calculations, make sure that you write neatly so you can read what you've written. Check to verify that the numbers or variables that you've written down are the same as those that appear in the question. Many incorrect answers are the result of carelessly changing information as you copy from the question to your work paper. Finally, work carefully as you do the necessary calculations to figure out the correct answer.

Step 3   **Double-check your answer.** A good way to do this is to retrace your steps and see if you come up with the same answer the second time around. In addition, make sure your answer is reasonable. For example, if you are asked to determine the speed of a race car, and your answer is 1,500 miles per hour, you must have made a mistake. No race car can travel that fast! Finally, you might try to estimate the correct answer by rounding the numbers that are provided in the question. For example, suppose you need to multiply 3.14 by 9 in order to figure out the circumference of a circle. You could round the numbers and multiply 3 by 10. Since 3 times 10 equals 30, your actual exact answer should be close to 30.

Step 4   **Read all the answer choices.** You may think the first or second answer choice is correct, and therefore you don't have to bother reading the other answer choices. It's a good idea to read **all** the answer choices to make sure the one you have chosen is correct. There may be a **better** answer in the choices you haven't read yet.

Step 5   **Use the process of elimination.** It is often easier to find the correct answer if you eliminate answer choices that you know are incorrect. If you're stumped by a question and you just can't figure out the correct answer, it's a good idea to identify the incorrect answer choices first. One or more of the answer choices may be completely unreasonable. Occasionally, some answer choices are just plain silly! If you see an answer choice that is obviously incorrect, you should eliminate it. Most tests allow you to write in your test booklet. In that case, you should cross out the answer choices that you eliminate.

If you can eliminate all the answer choices except one, that answer choice is probably correct. Even if you only knock out one or two answer choices, you will still increase your chances of answering the question correctly.

Step 6   **Fill in the bubble that corresponds to the question number and the answer choice you select.** This sounds easy. However, it is not uncommon for a student to bubble in letter *c* when he or she meant to bubble in letter *b*. Moreover, it is easy to mistakenly fill in the bubble for question 1 that you meant to fill in for question 2. This is especially true if you skip questions and plan to go back to them later.

Fair Wages for Students?

 In 1938, the Fair Labor Standards Act (FLSA) introduced the minimum wage. The first minimum wage was only twenty-five cents per hour. Of course, it has been raised many times since then. The current federal minimum wage was increased to $5.15 per hour in 1997. However, full-time students may be paid *less* than this. The FLSA allows for a minimum of $4.25 per hour for employees under twenty. This applies to the first ninety consecutive calendar days of employment. After ninety days, the employee must receive the actual minimum wage of $5.15 per hour. Although the law was meant to prevent abuse of child labor, many students think it is unfair.

▶ A student works part-time at a gas station. He earns $4.25 per hour plus tips. If you let $W =$ student's pay for the week, $h =$ the number of hours he works, and $t =$ tips he receives from customers, which of the following equations could be used to calculate his weekly pay?

A $\quad \$4.25t - h = W$
B $\quad \$4.25h = W + t$
C $\quad W = \$4.25h + t$
D $\quad W = \$4.25t + h$

Know It All Approach

Step 1 **Read the question carefully and notice the important information.** You are being asked to write an equation. Consider what information you need in order to write the equation. You'll need the amount of the hourly wage, $4.25. The variables W, h, and t are also important.

Step 2 **Answer the question.** The equation you write will be used to calculate the weekly pay. It should begin with "$W =$." Next, you need to figure out how the weekly pay is determined. An expression for this would be "$\$4.25h$." Add tips to the hourly wages by writing "$+ t$." Putting it all together, you come up with, "$W = \$4.25h + t$."

Step 3 **Double-check your answer.** A good way to check your answer is to plug in some numbers and see if the answer makes sense. For example, suppose Germaine worked 10 hours and received $12.00 in tips for a given week. According to your equation, his weekly pay would be

$$W = \$4.25h + t$$
$$W = (\$4.25)(10) + \$12.00$$
$$W = \$42.50 + \$12.00$$
$$W = \$54.50$$

Is this a reasonable answer? It makes sense. Anything less than $12.00 would be nonsense, and anything more than $100 would be excessive.

Step 4 **Read all of the answer choices.** Answer choice (C) is the same as your equation. Does this mean that (C) is correct? Probably. However, you should read all of the answer choices just in case you made a mistake.

Step 5 **Use the process of elimination.** By getting rid of answer choices that you *know* are incorrect, you will feel more certain you found the correct answer. Consider choice (A). The expression "$4.25t$" makes no sense. Why multiply the tips by the hourly wage? You can get rid of (A). Choice (B) seems to indicate that the hourly wage includes tips. This is incorrect, so get rid of (B). Choice (C) seems to be correct, so keep it. Look at choice (D). Again, the expression "$4.25t$" makes no sense. Get rid of choice (D). Choice (C) is correct.

Short-Answer Questions

Another type of question usually found on standardized tests is the short-answer variety. *Short-answer questions* ask you to write out your answer to a question. Unlike multiple-choice questions, the answer is not already printed on the paper for you to choose. You have to come up with the answer on your own. Fortunately, as the name implies, the answer is usually short. On most occasions, the answer is a number, a mathematical expression, or an equation. Here's the **Know It All Approach** to answering short-answer questions.

Step 1 **Read the question carefully.** As previously mentioned, this should *always* be the first step when answering *any* question on a test. Keep a sharp lookout for words such as **not, except, each, total, least, greatest, exact, approximate,** and **estimate.** Make sure you closely examine any diagrams, tables, charts, or graphs that accompany the question. They probably contain information needed to answer the question. Otherwise, they wouldn't be there!

Step 2

Answer the question. Again, this may seem obvious. However, since this type of question does not have a number of answer choices to consider, you must come up with the correct answer on your own. Write out the information you need to use, and then figure out the answer. Make sure that you have copied the information accurately.

Step 3

Double-check your answer. A good way to double-check your answer is to follow the same sequence of steps and answer the question a second time. Furthermore, it's a good idea to compare the information you used in your calculations with that provided in the question. If you copied a wrong number or a wrong formula, there is no way you will arrive at the correct answer.

Sometimes you can check your problem by reversing the steps you took to find your answer. If your calculations involve a subtraction problem, you can check your answer by addition. For example, suppose that in your calculations you subtract 27 from 100, and get an answer of 73. You can easily check this answer by adding 73 to 27. Their sum, 100, is the same as the number you started with, so it must be correct.

Step 4

Write your final answer in the space provided. For short-answer questions, you do not fill in a bubble. You need to *write out* your answer in the space provided on the answer sheet. Usually, the answer is a number, a mathematical expression, or an equation. Be sure to include a label for your answer (e.g., meters, square feet, miles, or degrees) if the question requires one.

Directions: Read the passage below. Then use the Know It All Approach to answer the sample problem on the next page.

The Shocking Story of Ranger Sullivan

In the United States, the odds of getting struck by lightning are one in 600,000. These are pretty long odds, but not long enough for an unlucky park ranger by the name of Roy C. Sullivan. Between 1942 and his death in 1983, Sullivan was struck by lightning, and lived to tell about it, seven times. Not that he escaped without a scratch. The first strike, in 1942, zapped off his big toenail. At various times throughout his run of poor luck, Sullivan's hair caught fire and his eyebrows were burned off. Sullivan's final encounter with lightning occurred in 1977. Sullivan survived the strike, but suffered burns on his chest and stomach.

▶ The probability of being struck by lightning is 1 in 125,000 or 0.000008. The probability of being left-handed is 1 in 8 or 0.125. What is the probability that one person would be both left-handed and struck by lightning?

Know It All Approach

Step 1 | **Read the question carefully.** The question is asking you to figure out the probability of a compound event. Notice the key word "both." This tells you that you are dealing with a compound event. A compound event consists of two or more simple events. That is, $P(A \text{ and } B) = P(A) \times P(B)$. Therefore, you need to multiply. Specifically, you need to consider the two probabilities—0.000008 and 0.125. You will find out more about probability in chapter 19 of this book.

Step 2 | **Answer the question.** Write out the information you need to answer the question. Make sure that you copy it accurately.

$P(A \text{ and } B) = P(A) \times P(B)$

$P(A \text{ and } B) = (0.000008)(0.125)$

$P(A \text{ and } B) = 0.000001$

Step 3 | **Double-check your answer.** Because this is a multiplication problem, you can use division to check your answer. 0.000001 divided by 0.125 equals 0.000008.

Step 4 | **Write your final answer in the space provided.** Usually, you are given a small space in which to write your answer. Because probability is a pure number, you do not need to label the answer with a unit. The correct answer is $P = 0.000001$.

Open-Response Questions

In recent years, standardized tests have placed less emphasis on multiple-choice and short-answer questions. They tend to ask a number of questions that require longer, more thought-out answers. *Open-response questions* ask you to write your answer to a question and show your work. By asking students to show all their work, those who grade the test have a better idea of the thought processes a student uses. Sometimes these questions have more than one part. Moreover, you can receive partial credit on these questions, even if you have the wrong answer. Here is the **Know It All Approach** to answering open-response questions.

Know It All Approach

Step 1

Read the question carefully. At the risk of being repetitious, this should **always** be the first step when answering a question. Reading carefully helps you understand what each question is asking you to do. Take notice of words such as **not, except, each, total, least, greatest, exact, approximate,** and **estimate.** Also, closely examine any numbers, variables, formulas, and equations that appear in the question. You will need to use them in order to figure out the answer.

Step 2

Use the space provided to calculate the answer. Write neatly and show all of your work. You should always show all of your work when answering open-response questions. If you are unable to get the correct answer, you may receive partial credit for showing your work and trying to answer it. Make sure your writing is neat and legible.

Step 3

Double-check your answer. A good way to recheck your answer is to retrace your steps and answer the question a second time. You should review the information provided in the question and compare it with the information you wrote down and used in your calculations. This is especially true for open-response questions because each step may be worth a few points toward your score.

Step 4

Make sure you answer all parts of the question. Write the answer or answers to the question very clearly. Some open-response questions include more than one part for you to answer. Be certain that you answer **all** parts of the question. Again, write your answers neatly and legibly, and show all your work.

Directions: Read the passage below and use the Know It All Approach method to answer the question that follows.

Unlucky Floor?

Some people refuse to stay on the thirteenth floor of a hotel or to work on the thirteenth floor of an office building because they think thirteen is an unlucky number. But, far from unlucky, the thirteenth floor is one of the safer floors in a building. The modern hook and ladder used by firefighters is equipped with a hydraulically operated crane mounted on a turntable. This equipment may be used in either fire fighting or rescue work. The crane can be extended up to 150 feet in length. The maximum *vertical* height the crane can reach is usually no more than 130 feet. This is the equivalent of about thirteen stories or floors. From a safety perspective, the thirteenth floor is lucky indeed!

▶ The extension ladder of a fire truck is 150 ft. long (side *c*). The base of the ladder is 90 ft. from the side of the building (side *b*). According to the Pythagorean theorem, how far up the side of the building (side *a*) does the ladder reach? **Show all work.**

Angle ACB = 90°   The Pythagorean Theorem: $c^2 = a^2 + b^2$

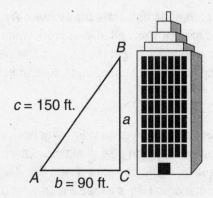

Know It All Approach

Step 1   **Read the question carefully.** The question asks you to figure out the vertical height (side *a*) reached by the tip of the ladder. You should note that triangle ABC is a right triangle, with hypotenuse AB and legs AC and BC. Line AB = side *c* = 150 feet; and line AC = side *b* = 90 feet. The question also provides you with a formula, that is, the Pythagorean theorem.

Step 2 **Use the space provided to calculate the answer. Write neatly and show all your work.** The first step should be to write the formula. You should realize that a is the side opposite angle A, b is the side opposite angle B, and c is the side opposite angle C.

$$c^2 = a^2 + b^2$$

Next, substitute the information given in the question.

$$c(150)2 = a^2 + (90)^2$$

Finally, do your calculations and solve for a.

$$22{,}500 = a^2 + 8{,}100$$
$$14{,}400 = a^2$$
$$120 = a$$

Draw a box around your final answer, and label it with the proper unit.

$$\boxed{a = 120 \text{ ft.}}$$

Step 3 **Double-check your answer.** A good way to recheck your answer is to plug it into the equation and see if it works.

$$c^2 = a^2 + b^2$$
$$(150)^2 = (120)^2 + (90)^2$$
$$22{,}500 = 14{,}400 + 8{,}100$$
$$22{,}500 = 22{,}500$$

Because the plugged in values worked out, you know you found the correct answer.

Step 4 **Make sure you answer all parts of the question. Write the answer or answers to the question very clearly.** Although this question has only one part, it requires you to show all your work. Make sure you have included all the steps you followed in figuring out the answer.

Subject Review

Now you know what sort of questions you can expect to find on a math test, and you've found the answers to the three questions at the beginning of the chapter.

Why are working teenagers paid less than the minimum wage?

The Fair Labor Standards Act permits employers to pay a minimum wage of $4.25 per hour for employees less than twenty years of age.

What are the chances of being struck by lightning?

One in 125,000. But try telling that to Ranger Sullivan!

Why is the thirteenth floor of a building a "lucky" floor?

You could consider the thirteenth floor "lucky" because fire truck rescue ladders can only reach as high as the thirteenth floor.

Chapter 2: Computation

How many new red blood cells does the human body make every second?

When will Los Angeles slide past San Francisco?

How many dozens of donuts does Dunkin' Donuts sell every year?

Integers

Integers are the set of whole numbers and their opposites. Examples of integers are −3, 8, −5, 0, and 4. Positive integers may be shown with or without the plus (+) sign. Negative integers are always shown with a minus (−) sign. Zero is the only integer that is neither positive nor negative.

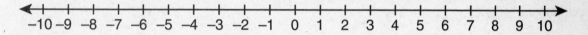

You can use a number line to visualize integers. Numbers on a number line increase from left to right. The least number is farthest to the left, and the greatest number is farthest to the right. For example, −3 is greater than −4 because it is to the right of it on the number line. The opposite counterpart of an integer (e.g., +5) is the same distance from zero, except it is on the opposite side (−5).

Adding integers is straightforward. One approach is to pretend you are figuring out a budget. Use positive integers to represent money that is coming in, such as salary and tips. Use negative integers to represent money that is going out, such as bills and debts to be paid. If you make $500 in a week, and you have to pay $350 in bills, the net result is $500 + (−$350), or +$150. If, in the next week, you make $560, and you have to pay $780 in bills, the net result is $560 + (−$780), or −$220.

There's no trick to **subtracting integers** either. Think of it as *adding the opposite* of a number. If you make $450 in a week, and you also get rid of $600 in debt, the net result is $450 − (−$600) = $450 + $600, or +$1,150.

Multiplying and dividing integers is a bit more complicated. If two integers have the *same* sign, the product or quotient is positive. If two integers have *different* signs, the product or quotient is negative.

$2 \times 5 = 10$ $\qquad$ $-2 \times -5 = 10$ $\qquad$ $-2 \times 5 = -10$ $\qquad$ $2 \times -5 = -10$

Absolute Value

The **absolute value** of a number is the number of units a number is from zero on the number line. It is the value of a number without regard to the number's sign. For example, on the number line shown below, which letter represents the number with the greatest absolute value?

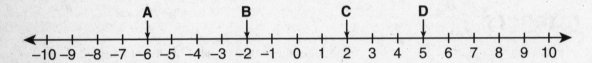

You may have been tempted to guess **D** because it is farthest to the right. However, the correct answer is **A** because it is the greatest number of units from zero. The absolute value of -6 can be written as $|-6|$.

$$|-6| = 6$$

Imagine that the result of a vote on a motion in the U.S. Senate is 40 for and 60 against. Think of the numbers as $+40$ and -60. The motion would be defeated because $|-60|$ is greater than $|+40|$.

Order of Operations

Sometimes math computations can get very complicated. For example, imagine that you are asked to simplify the following expression: $19 - 2 \bullet 5^2 + (35 - 3) \div 4$.

Where do you begin? Mathematical expressions may contain parentheses, exponents, operational signs, and variables. To help you decide what to do first, mathematicians came up with the "order of operations."

$19 - 2 \cdot 5^2 + (35 - 3) \div 4$	P: Do all work in **Parentheses** first.
$19 - 2 \cdot 5^2 + 32 \div 4$	E: Next, simplify all expressions with **Exponents**.
$19 - 2 \cdot 25 + 32 \div 4$	M: **Multiply** and
$19 - 50 + 8$	D: **Divide** in order from left to right.
	A: **Add** and
-23	S: **Subtract** in order from left to right.

Directions: Read the passage below and answer the question that follows.

The Most Important Organ

Your heart is a very special organ. It beats about 100,000 times a day. About nine thousand quarts of blood move through it every day. Your heart never takes a break. During your lifetime, it will beat about three billion times and pump about four hundred million quarts of blood. Blood is carried away from the heart by arteries, and is returned to the heart by veins. However, most of your blood is located in the capillaries, the tiniest blood vessels. Capillaries are so small that red blood cells can only pass through them in single file! The human body contains about sixty thousand miles of capillaries. That is enough to reach around Earth's equator nearly two-and-a-half times. Your blood is very special too. It contains platelets to help your blood clot, white cells to fight germs, and red cells to carry oxygen. The human body makes about two million red blood cells every second!

▶ If the human heart beats about 100,000 times every day, about how many times does it beat every minute? **Show all work.**

_____

Know It All Approach

First, read the question carefully and make note of the important words and information. The terms "every day" and "every minute," and the number "100,000" are important in order to answer the question. Take special notice of the word "about" in the question. This means you can round your answers at each stage to simplify your calculations.

Once you've gathered up all the information you need, calculate the answer. Because this is an open-response question, you must show all of your work.

The first thing you need to do is figure out how many times the human heart beats every hour. Since there are twenty-four hours in a day, you need to divide 100,000 by 24.

$$100,000 \div 24 = 4,167 \text{ (rounded) beats per hour}$$

Now that you have figured out how many times the human heart beats every hour, the next step is to figure out how many times it beats every minute. Because there are sixty minutes in an hour, you need to divide 4,167 by 60.

$$4,167 \div 60 = 69.45 \text{ or about 69 beats per minute}$$

When you've finished your calculations, check your answer. You used division to find the answer, so a good way to double-check it would be to multiply.

$$60 \times 69 = 4,140 \qquad 24 \times 4,167 = 100,008$$

This is quite close to the original number given in the problem. It is slightly different only because you rounded your answers.

If this was a test situation, you should make sure you answer all parts of the question. Write the answer or answers to the question very clearly.

Directions: Read the passage below and answer the questions that follow.

Cities on the Go

Today, Los Angeles, California, is about 240 miles southeast of San Francisco. However, most geologists believe that this will not always be the case. They have concluded that each year, these two cities move 2.5 inches closer to each other because they are on opposite sides of the San Andreas Fault. The theory of plate tectonics proposes that changes in Earth's crust are caused by the very slow movement of large crustal plates. The San Andreas Fault is a boundary where two crustal plates are grinding past each other. At this rate, Los Angeles will slip past San Francisco in approximately 31.5 million years.

1. If the Pacific Plate is moving at a rate of 2.5 inches per year, about how many years will it take to move one mile? (1 inch ≅ 0.00002 miles.)

   A   5,000
   B   10,000
   C   2,000
   D   50,000

2. If you multiply −3 by an integer less than −1, which of the following will be the result?

   A   an integer less than −3
   B   an integer between −3 and 3
   C   an integer less than 3
   D   an integer greater than 3

3. Simplify the following: $|5 - 9| - |-2|$.

   A -6
   B -2
   C   2
   D   6

4. Simplify the following: $2 \cdot 6^2 \div 3 + (4^3 - 55)$.

   A   6
   B   9
   C   24
   D   33

5. What is the solution set for y, if $|y| = 4$?

   A $\{-4, 0\}$
   B $\{0, 4\}$
   C $\{-4, 4\}$
   D $\{-16, 16\}$

The Land of Excess

The United States is often referred to as "The Land of Plenty." Some might argue that "The Land of Excess" would be a more accurate description. On a given day, Americans eat more than three hundred million slices of pizza. On average, each American consumes approximately 260 pounds of meat in a year. In the year 2000, Americans consumed more than twenty billion hot dogs. The average American will eat nineteen pounds of cereal and 135 pounds of sugar per year. Here's the clincher. Every year, Dunkin' Donuts sells about 2.4 billion donuts!

6. If Dunkin' Donuts sells 2.4 billion donuts every year, how many dozens of donuts does it sell per year?

_____

7. If Americans eat 302,400,000 slices of pizza every day, how many slices do they eat every second? **Show all work.**

_____

Subject Review

In Chapter 2, you learned how to perform calculations with integers. Integers are whole numbers and their opposites. You also learned that the absolute value of a number depends on how far the number is from zero, regardless of whether it is positive or negative. Finally, you learned how to simplify mathematical expressions. The steps you need to follow are given in the order of operations.

You also learned the answers to some interesting questions.

How many new red blood cells does the human body make every second?
The human body makes about two million new red blood cells a second.

When will Los Angeles slide past San Francisco?
Some scientists predict this will happen in about 31.5 million years.

How many dozens of donuts does Dunkin' Donuts sell every year?
About two hundred million dozens are sold in Dunkin' Donuts each year.

Chapter 3: Word Problems

What is the longest elevator fall in which all the passengers survived?

What animal is the world's fastest distance runner?

What is the world's largest sculpture?

Word Problems

A **word problem** uses words to state a problem, but it requires mathematics to solve it. Examples such as "Simplify: $8^2 - 5 \bullet 6$" are straightforward. You know by looking at it that you need to square the 8, multiply the 5 times the 6, and then subtract.

$$8^2 - 5 \bullet 6$$
$$64 - 30 = 34$$

Some word problems are more complicated. They give you the information in words and numbers, but they do NOT tell you what to do with those numbers. You have to figure that out on your own. If you read the question carefully, the words will tell you which operations you need to perform. In addition, some word problems require more than one step to find the answer. Consider the following problem.

▶ The basketball coach is ordering new uniforms. Each of the 15 students on the team will get a new tank top ($19 each), a new pair of shorts ($16 each), and a new pair of basketball shoes ($59 each). What is the **total** cost of the new uniforms for **all** the students on the team?

The problem presents you with four different numbers (15, $19, $16, and $59). However, it does not tell you whether to add, subtract, multiply, or divide. By reading the question carefully, you should be able to figure out what needs to be done.

What does the question ask you to find out? *The total cost*

What do you need to do? *Multiply the cost of each item by 15.*

What is the last step? *Add the total costs together.*

tank tops: $15 \bullet \$19 = \$285$

shorts: $15 \bullet \$16 = \$240$

shoes: $15 \bullet \$59 = \underline{\$885}$

Total Cost = $1,410

Rate

Many math word problems are related to the concept of rate. **Rate** is a ratio that compares quantities measured in different units. Rate of speed is measured in miles per hour; heart rate in beats per minute; the cost of eggs in price per dozen; and wages in pay per hour. These are all examples of rate. Try this problem.

▶ Henry ran the marathon (about 26 miles) in 2.6 hours. What was his average rate of speed?

Distance = rate × time

$26 = s \times 2.6$

$s = 26 \div 2.6$

$s = 10$ miles per hour

Another type of rate is **rate of change.** Imagine that a sports car can go from 0 to 60 miles per hour (mph) in 4 seconds. What is its rate of change of speed (a = acceleration)?

a = change of speed ÷ time for that change

$a = (60 - 0) \div 4$

$a = 15$ mph per sec.

Directions: Read the passage below and answer the question that follows.

The Ride of Their Lives

It happened on January 25, 2000. Shameka Peterson and Joe Mascora worked on the forty-fourth floor of the famed Empire State Building in New York City. They boarded one of the building's sixty-four elevators to go down to street level. When the elevator cable snapped, they fell forty stories (four hundred feet) in four seconds! The emergency braking system brought them to an abrupt halt just four stories above the ground. Both suffered only minor injuries. Following an investigation, the New York City Council reported that from 1992 to 1996, 369 elevator accidents had occurred. This averages out to about one accident every five days.

▶ Imagine that 425 elevator accidents occurred from 1997 to 2003. That would average out to about one accident every how many days?

   A   every 3 days
   B   every 4 days
   C   every 5 days
   D   every 6 days

First, read the question carefully and find the information you need to solve the problem. The numbers "425" and "1997 to 2003" are important, as are the words "about one accident every how many days." Notice the word "about." This means you are looking for an estimate.

Next, solve the problem. The time period from 1997 to 2003 covers seven years. If you multiply 7 × 365 = 2,555, you get the number of days in seven years. If there were 425 accidents over the course of 2,555 days, you can figure out the average by dividing 2,555 by 425.

$$2{,}555 \div 425 = 6.011 = \text{about one accident every 6 days}$$

Then, check your answer. A good way to double-check your answer is to use the reverse operation. Divide 2,555 ÷ 365 = 7. That is the number of years given in the problem. Next, multiply 425 × 6 = 2,550. That is close to the number of days in 7 years. Remember, you rounded your answer.

After you have checked you math, read all the answers. Only answer choice (D) agrees with your answer.

Directions: Read the passage below and answer the questions that follow.

On the Fast Track

Peregrine falcons are the fastest animals in the world. They can achieve speeds approaching 124 mph when plunging from the sky after prey. One has been clocked by radar at 114 mph after a dive of 305 meters (1,000 ft.). Of course, diving through the air is quite different from running on land. Humans can run about 21 mph. The fastest dog, the greyhound, can reach speeds of up to 41.7 mph. An ostrich can run up to 43 mph. However, the world's fastest mammal, the cheetah, runs at a maximum speed of 63 mph. Remarkable also is its rate of acceleration—from a standing start to a speed of 45 mph in just two seconds. However, it can maintain its top speed only for about 300 yards before becoming exhausted. What animal is the world's fastest distance runner? That honor goes to the pronghorn, or American antelope. It can run continuously at a speed of 42 mph for one whole mile. What's more, it can maintain a speed of 35 mph for 4 miles.

1. How much time would it take for an American antelope, running at an average speed of 30 miles per hour, to cover a distance of 6 miles?

   A   12 minutes
   B   30 minutes
   C   2 hours
   D   5 hours

2. Alicia is planning to run in a 10K (10 km) race three weeks from now. She trains by running 5 km each Tuesday, Thursday, and Saturday. On each Monday and Friday, she runs 15 km. She rests on Wednesday and Sunday. After three weeks of training, Alicia will have run the equivalent of how many 10K races? **Show all work.**

3. In 1975, the tuition at Einstein University was $16,000 a year. By 1995, the tuition had increased by 25%. If tuition continues to increase at this rate, what will be the tuition in 2005?

   A $4,000
   B $20,000
   C $25,000
   D $31,250

Directions: Read the passage below and answer the questions that follow.

Mount Rushmore

South Dakota's Black Hills provide the backdrop for Mount Rushmore, the world's greatest mountain carving. This epic sculpture features the faces of four American presidents: George Washington, Thomas Jefferson, Abraham Lincoln, and Theodore Roosevelt. Gutzon Borglum, the chief sculptor behind the project, began drilling into the 5,725-foot mountain in 1927. The project took 14 years and cost $1 million. Only about six and a half years were spent on actual carving. The rest of the time was lost due to weather delays and lack of funding. The process of carving was difficult and dangerous. Dynamite was used to blast the rock into rough shapes. After blasting, workers would be lowered down the face of the mountain using cables and swing chairs to carve out the fine features with hammers and chisels. Work continued on the project until Borglum's death in 1941. The sculpture is unfinished! The original plan was to carve details of the presidents down to the waist. However, no carving has been done on the mountain since that time and none is planned in the future.

4. At Mount Rushmore, the presidents' heads are each 60 feet tall. If the average person's face is 0.75 feet (9 inches) high, and each of the presidents was about 6 feet tall, how tall would a sculpture of each president's entire body be?

_____

5. The head of the school's woodshop ordered two kinds of drill bits. He paid $84 for 48 steel drill bits and $78 for 24 cobalt drill bits. How much more did each cobalt drill bit cost than each steel drill bit?

    A $0.25
    B $1.50
    C $3.00
    D $6.00

Subject Review

Now you know how to tackle any word problems a test may throw your way. First, read the question carefully and pick out any information you'll need and any clue words that will let you know what operations you must perform. Then, solve the question and check your work. Finally, read all the answer choices and pick the best answer choice. Don't forget to use the process of elimination, and pick out the answers you know are wrong, if you have trouble finding the correct answer.

Furthermore, you know enough to answer the three questions at the beginning of the chapter.

What is the longest elevator fall in which all the passengers survived?
Two passengers fell 400 ft.

What animal is the world's fastest distance runner?
The pronghorn, or American antelope can run over 40 mph. The cheetah is faster, but only for short distances.

What is the world's largest sculpture?
Mount Rushmore in South Dakota is the world's largest sculpture, although the nearby Crazy House sculpture is taller.

Chapter 4: Fractions, Decimals, and Percents

Why are sneaker companies talking to LeBron James?

What makes up the major part of a dust mite's diet?

How long does it take to jump twenty-three miles on a pogo stick?

Fractions, Decimals, and Percents

A **fraction** is a number that names part of a whole. For example, if a pizza pie is cut into eight slices, and you eat three of them, then you have eaten $\frac{3}{8}$ of the whole pie.

A **decimal** is a number written with a decimal point. For example, the amount of rainfall in a given month could be expressed as 0.75 inches. This number is read as "seventy-five hundredths." Another way of writing it is "$\frac{75}{100}$." So you see, a decimal is actually a fraction whose denominator is a power of 10.

A **percent** is a ratio that compares a number to 100. For example, if 15% of the automobiles sold are manufactured by Japanese companies, then fifteen out of every one hundred cars are made in Japan. The term "percent" means literally "per hundred." So you can think of a percent as a fraction whose denominator is always 100. Hence, 15% equals $\frac{15}{100}$. Try the sample question below.

▶ List the following numbers in order from smallest to largest: $\frac{1}{2}$, -0.4, 20%, $-\frac{4}{5}$, 0.75.

How do you solve a problem like this? You know that the negative numbers are smaller than the positive numbers. What is the next step? In order to find the correct answer, you need to be able to convert among fractions, decimals, and percents.

Converting Fractions, Decimals, and Percents

One way to convert a fraction to a decimal is to find an equivalent fraction whose denominator is a power of 10.

$$\frac{2}{5} \times \frac{2}{2} = \frac{4}{10}, \text{ or } 0.4.$$

Another way to convert a fraction to a decimal is to divide the denominator into the numerator.

$$\frac{3}{8} = 8\overline{)3} = 0.375$$

Converting a decimal to a fraction is even easier. Simply write the decimal as a fraction. Then reduce it to lowest terms.

$$0.15 = \text{fifteen hundredths} = \frac{15}{100} = 15 \div \frac{5}{100} \div 5 = \frac{3}{20}$$

Converting a decimal to a percent is also easy. All you have to do is move the decimal point two places to the right and add the percent sign.

$$0.37 = 37\%$$

Similarly, you can convert a percent to a decimal by reversing the process. Take away the percent sign and move the decimal point two places to the left.

$$79\% = 0.79$$

Now return to the sample problem. The best way to find the answer is to convert all of the numbers to decimals. Two of the terms are already in decimal form: -0.4 and 0.75. You will have to convert the rest.

$$\frac{1}{2} = 0.5 \qquad 20\% = 0.20 \qquad -\frac{4}{5} = -0.8$$

Now it's easy to put the numbers in order from smallest to largest. See how they are plotted on the number line below.

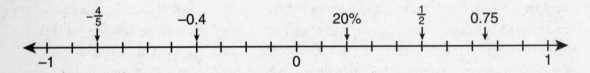

Computing with Fractions

▶ Add the following fractions: $\frac{1}{3} + \frac{1}{8}$

Before you can add fractions, you have to change them to equivalent fractions that have the same denominator. This is also true for subtraction.

$$\frac{1}{3} = \frac{1}{3} \times \frac{8}{8} = \frac{8}{24} \qquad \frac{1}{8} = \frac{1}{8} \times \frac{3}{3} = \frac{3}{24}$$

$$\frac{8}{24} + \frac{3}{24} = \frac{11}{24}$$

▶ Felipe delivers newspapers after school. His friend, Paula, helped him deliver the papers last week. Felipe earns $\frac{1}{8}$ of all the money he collects. He wants to give Paula $\frac{1}{2}$ of what he earns. What fraction of the money that Felipe collects will he give to Paula?

This question is really asking, "What is $\frac{1}{2} \times \frac{1}{8}$?" Multiplication of fractions is easy. To find your answer, multiply the numerators and then multiply the denominators.

$$\frac{1}{2} \times \frac{1}{8} = \frac{1 \times 1}{2 \times 8} = \frac{1}{16}$$

To divide fractions, you *invert* the divisor and then multiply.

$$\frac{5}{9} \div \frac{2}{3} = \frac{5}{9} \times \frac{3}{2} = \frac{5 \times 3}{9 \times 2} = \frac{15}{18} = \frac{5}{6}$$

Computing with Decimals

▶ The school soccer field measures 86.25 meters long by 44.8 meters wide. The coach wants to divide the field into four equal-size rectangles to practice various drills. How many square meters would each of the four smaller rectangles take up?

To find the answer to this problem, you need to calculate the area of the soccer field, and then divide by four. The formula for the area of a rectangle is: Area = length × width. (You will learn more about area in Chapter 12.)

Area = length × width = 86.25 × 44.8 = 3,864 square meters

3,864 ÷ 4 = 966 square meters = area of each rectangle

▶ A pack containing six 12-ounce soda cans costs $2.69. How much does each can cost? (Round to the nearest cent.)

A $0.40
B $0.44
C $0.45
D $0.48

The United State's monetary system is based on powers of ten. Therefore, you can treat any question that refers to dollars and cents as a decimal problem. In this problem, you need to divide $2.69 by 6.

$2.69 ÷ 6 = 0.4483 = $0.45 rounded to the nearest cent. Therefore, choice (C) is correct.

Computing with Percents

▶ A company makes inkjet cartridges. Typically, after inspection, 5% of the cartridges it makes are rejected as defective. If the company made 1,240 cartridges last week, how many good cartridges will be left following inspection?

A 62
B 620
C 1,178
D 1,235

To answer this problem, you need to find 5% of 1,240, and then subtract that number from 1,240. Before you can multiply 1,240 by 5%, you must first convert 5% to either a fraction or a decimal. Recall that, to convert a percent to decimal, you should take away the percent sign and move the decimal point two places to the left. 5% = 0.05.

$1,240 \times 0.05 = 62$ = the number of cartridges rejected

$1,240 - 62 = 1,178$ = the number of good cartridges left

Therefore, choice (C) is correct. If you had converted 5% to a fraction, you would get the same answer. Check it out below.

$5\% = 0.05 = \dfrac{5}{100} = \dfrac{1}{20}$ $1,240 \times \dfrac{1}{20} = 62$

▶ If 27 students in a class of 30 pass a science test, what percent of the class passed?

A 10%
B 27%
C 81%
D 90%

To solve this one, you have to change the fraction, $\dfrac{27}{30}$, to a percent.

$\dfrac{27}{30} = \dfrac{9}{10} = 0.9 = 90\%.$

Choice (D) is correct.

► Last season, a ticket for a hockey game cost $36.00. This season, the same ticket costs $45.00. What is the percent increase in price?

A   20%
B   25%
C   45%
D   80%

Questions involving percent increase or decrease can be tricky. The most important thing to remember is that the answer is the percent change of the original number *before* the increase or decrease. In this question, you need to subtract to find the amount of increase. Then, you need to find what percent that amount is of $36.00 (the original number).

$$45.00 - 36.00 = 9.00 = \text{amount of increase}$$

$$\frac{9.00}{36.00} = \frac{9}{36} = \frac{1}{4} = 0.25 = 25\%$$

Choice (B) is correct. Some students might mistakenly find what percent $9.00 is of $45.00 (20%). That would lead them to choose (A) as the correct answer. However, $45.00 was *not* the original number before the increase and choice (A) would be incorrect.

► Allen put $500.00 in a special bank account. It pays 4% simple interest each year. Allen keeps the money in the bank for three years. How much interest will Allen earn in three years?

A $12.00
B $20.00
C $60.00
D $200.00

This is an interest problem. Perhaps you remember the formula, $I = P \times R \times t$, that is, Interest = Principal × Rate × time. The principal is $500.00. The rate is 4%. The time is 3 years.

$$I = P \times R \times t$$

$$I = 500.00 \times 0.04 \times 3$$

$$I = \$60.00$$

The correct choice is (C).

LeBron James Got Game

Imagine that you are eighteen years old, and the whole world is watching you. You play basketball so well that your home games are shown on pay-per-view! There has been so much hype because you will be a Number One draft pick in the National Basketball Association (NBA) draft. Sneaker companies are ready to pay you millions of dollars to endorse their athletic shoes. Do you think you could handle it? This is the exaggerated lifestyle of LeBron James. The 6'8", 240 pound senior played basketball at St. Vincent-St. Mary High School in Akron, Ohio. The pro scouts were salivating at the prospect of signing him to a multiyear, multimillion dollar contract. They described him as a fundamentally sound player with excellent form on his jump shot, brilliant passing skills, great leaping ability, and dazzling finishes around the rim. He has become one of the young stars of the NBA, going pro at the age of eighteen. "People ask me if it's a hard decision going to the NBA, but I've made harder decisions," he says. "Decisions about smoking or going to school, or stealing from a store or not stealing. Those are harder decisions."

▶ LeBron James's free throw shooting percentage is 60%. If he took 15 free throws in a game, how many would you expect him to make?

First, read the questions carefully. Make sure you understand what the question is asking and find all the information you'll need to calculate your answer. The term "free throw shooting percentage," and the numbers "60%" and "15" are important in order to answer the question. You need to find 60% of 15.

Now that you know what you're looking for and you have the necessary information, calculate the answer.

$$60 \times 15 = 0.60 \times 15 = 9$$

After you've calculated your work, check your answer to be sure you're right. Because you used multiplication to find the answer, a good way to double-check it would be to divide.

$$9 \div 15 = 0.60 = 60\%$$

Finally, because this is a short-answer question, make sure you have answered all parts of the question (in this case there's only one part, so that's covered), and be sure to write your answer clearly in the space provided.

Directions: Read the passage below and answer the questions that follow.

Microscopic Monster

A person sheds about fifty million dead skin cells every day. Ninety percent of dust particles in your home are from dead skin. Dust mites primarily feed on dead skin cells regularly shed from humans and their animal pets. They love warm, humid areas filled with dust. Sofas, chairs, carpets, bed pillows, and mattresses are common dust mite homes. A typical bed usually houses six billion dust mites. They are about $\frac{1}{4}$ mm (0.01 inches) long, and thus invisible to the naked eye. For most people, dust mites are not harmful. However, some people develop severe allergies to dust-mite *droppings*. If you lie on a rug where they live, you might get itchy red bumps on your skin. If you breathe in dust, you may have difficulty breathing or even a severe asthma attack.

So cutting down the number of dust mites in the home is an important step if you have allergies or asthma. You should thoroughly vacuum all furniture, carpeting, and mattresses. In addition, you should change bed linens often.

1. Scientists are testing a new spray that is designed to eliminate dust mites. A sample estimated to contain 72 million live dust mites was treated with the spray. After a few hours, the estimated number of live dust mites in the sample was 27 million. What was the percent decrease in the number of live dust mites?

   A 37.5%
   B 45.0%
   C 62.5%
   D 99.0%

2. List the following numbers in order from smallest to largest: $\frac{1}{3}$, 1.3, 13%.

   A $\frac{1}{3}$, 1.3, 13%

   B 1.3, 13%, $\frac{1}{3}$

   C $\frac{1}{3}$, 13%, 1.3

   D 13%, $\frac{1}{3}$, 1.3

3. Simplify: $\frac{14}{15} - (\frac{1}{3} + \frac{1}{5})$

   A $\frac{2}{5}$

   B $\frac{2}{3}$

   C $\frac{4}{5}$

   D $\frac{12}{7}$

Directions: Read the passage below and answer the questions that follow.

Pogo On the Go

On June 27, 1997, the Queensborough Community College running track was the site of a truly unique record-breaking feat. On that day, Ashrita Furman of Jamaica, New York, set a pogo stick jumping distance record of 37.18 kilometers (23.11 miles). Furman jumped around the track for more than 12 hours on his four-foot long pogo stick. Nobody was counting individual jumps, so there is no record of how many jumps were required to break the record. Though this record would be enough to ensure Ashrita Furman an honored place in this *Know It All!* book, Furman gets special recognition for holding more Guinness World Records than any individual. His other records include such strange stunts as "most milk crates balanced on a chin," "most hopscotch games played in a twenty-four hour period," and "most underwater rope jumps."

4. If 37.18 kilometers is equal to 23.11 miles, then, to the nearest tenth, how many kilometers are in one mile?

    A   0.6
    B   1.6
    C   6.0
    D   14.1

5. The nutrition label on a can of tomato sauce indicates that each serving contains 380 milligrams of sodium. The company recently came out with a low-salt tomato sauce that contains only 15 milligrams of sodium per serving. What is the percent decrease in the number of milligrams of sodium per serving of regular tomato sauce compared with the low-salt variety?

    _____

Subject Review

So now you know the difference between a decimal, a fraction, and a percent. A fraction is a number that names part of a whole and is written by placing one number on top of another, like so: $\frac{2}{3}$. A decimal is a number written with a decimal point. A percent is a ratio that compares a number to 100. You also know how to compare decimals, fractions, and percents by converting them.

You practiced answering questions that required you to calculate fractions, decimals, and percents.

Last but not least, you can also answer the three questions at the beginning of the chapter.

Why are sneaker companies talking to LeBron James?
They want him to endorse their sneakers. Nike paid James $90 million to pitch its footwear.

What makes up the major part of a dust mite's diet?
Dust mites eat dead skin cells.

How long does it take to jump twenty-three miles on a pogo stick?
It took Ashrita Furman about $12\frac{1}{2}$ hours.

Chapter 5: Ratio and Proportion

What is the fastest-swimming aquatic mammal?

What country requires school-age children to wear hats and long-sleeved shirts when they have gym class outdoors?

How did Rosie Ruiz manage to sneak to the front of the line in the 1980 Boston Marathon?

Ratio and Proportion

A **ratio** is a comparison of two numbers by division. For example, suppose a history class has eighteen girls and twelve boys. The ratio of girls to boys is 18 to 12 or, reduced to lowest terms, 3 to 2. A ratio may be expressed in words (three to two), as a fraction $\left(\frac{3}{2}\right)$, or with a symbol (3:2). Study the sample problem below.

▶   During a blackout, a hardware store sold thirty-two packets of size AA batteries and its whole supply of packets of size D batteries. If the ratio of packets of size D batteries sold to packets of size AA batteries sold was 15 to 8, how many packets of size D batteries were sold?

The question states that the ratio of size D packets sold to size AA packets sold is 15 to 8. In addition, you are told that thirty-two packets of size AA batteries were sold. You need to figure out how many packets of size D batteries were sold.

The easiest way to figure this out is to set up a proportion. A **proportion** is an equation that shows that two ratios are equivalent. For example, consider again the history class. You can set up the following proportion.

$$\frac{18}{12} = \frac{3}{2}$$

The ratio on the left is equivalent to the ratio on the right. Now use this strategy to answer the question.

$$\frac{\text{size D}}{\text{size AA}} = \frac{15}{8} = \frac{x}{32}$$

You set up a proportion, letting x equal the number of packets of size D batteries sold. If you know three of the four numbers in a proportion, you can cross multiply to find the unknown number.

$$\frac{15}{8} = \frac{x}{32}$$

$$8 \times x = 15 \times 32$$

$$8x = 480$$

$$x = 60$$

The answer is sixty packets of size D batteries were sold. You can check to make sure the answer is correct. Is the ratio 60 to 32 the same as the ratio 15 to 8? Yes it is! Therefore, the answer is correct.

Scale

You have undoubtedly seen a scale model in your lifetime. Some people build and collect model airplanes or model cars. A scale model is a replica of an original object that is too large to be built at actual size. Perhaps you have seen a scale drawing, such as the floor plan of a house. A scale drawing is a drawing that is similar to but smaller than the actual object.

Scale is the ratio of a given length on a drawing or model to its corresponding length in reality. If you know the scale used for a model or drawing, you can determine the actual measurement. Try this sample problem.

▶ The model of the Space Shuttle shown above is twenty-three inches long, and the model was made using a scale of 1:64. How long is the *real* Space Shuttle?

The scale tells you that one inch on the model represents sixty-four on the real shuttle. If the model is twenty-three inches long, then the real shuttle is 23 × 64, or 1,472 in. long. It is difficult to picture that many inches in your mind, so it might be a good idea to convert your answer from inches to feet. Since 12 inches = 1 foot, then 1,472 inches = 122 feet (1,472 ÷ 12). Therefore, the *real* Space Shuttle is 122 feet, 8 inches long.

Here's another sample problem.

▶ The other diagram shown above is an architect's scale drawing of the floor plan of a house. The drawing measures $5\frac{1}{4}$ inches by $4\frac{3}{4}$ inches. If the actual house measures 63 feet by 57 feet, by what factor should you multiply all of the measurements?

A 12
B 48
C 64
D 144

Some questions may require you to measure a scale distance using a ruler. To convert the scale distance to actual distance, you must use a distance scale, such as a scale of miles, provided with the diagram.

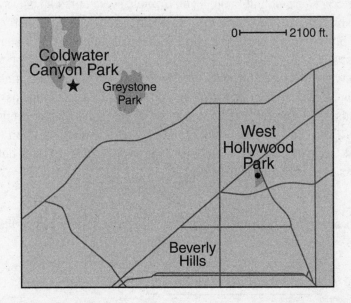

▶ The diagram above is a map of Beverly Hills, CA, 90210 (the zipcode made famous by a television show). A distance scale is located in the upper-right corner. What is the approximate distance between Coldwater Canyon Park (the star) and West Hollywood Park?

Use a ruler to measure the scale distance between the two points given in the question. Your measurement should be about two inches. Next, line up the zero line on your ruler with the zero line on the distance scale. Notice that one-half inch on the distance scale is equivalent to an actual distance of 2,100 feet. Finally, set up a proportion, letting x stand for the actual distance between the two parks.

$$\frac{0.5 \text{ in.}}{2{,}100 \text{ ft.}} = \frac{2 \text{ in.}}{x}$$

$$0.5x = 2 \times 2{,}100$$

$$0.5x = 4{,}200$$

$$x = 8{,}400 \text{ ft.}$$

Directions: Read the passage below and answer the question that begins below it.

Nature's Speed Swimmers

What are the chances that you could out-swim a shark? Not very good! Sharks can reach speeds of 88 kilometers per hour (kph). That's equivalent to about 55 miles per hour (mph). Even Olympic swimmers can only swim as fast as 8 kph (5 mph). The fastest swimmer is the sailfish, which can reach 109 kph (68 mph). The orca, or killer whale, is the fastest aquatic mammal. It can swim at a speed of 48 kph (30 mph). This may surprise you, but a hulking polar bear can swim as fast as 10 kph (6 mph). Dolphins usually attain speeds of 35 kph (22 mph). Seals swim at about 19 kph (12 mph), and penguins can reach speeds of 24 kph (15 mph). Do you recall the story of the tortoise and the hare? Even the slow but steady sea turtle can swim faster than a human— about 9 kph (6 mph). The slowest swimming fish is the sea horse, which moves at about 0.16 kph (0.01 mph).

▶ On a graph, a drawing of a killer whale is placed next to a square ceiling tile to show their relative sizes.

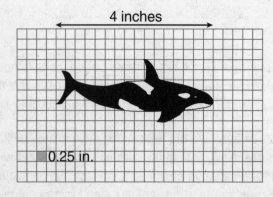

4 inches

0.25 in.

Scale
□ = 0.25 inch

The image of the square ceiling tile is 0.25 inches on each side, and the image of the killer whale is 4 inches long. If the actual ceiling tile measures 24 inches on each side, what is the actual length (in feet) of the killer whale?

A   2 ft.

B   8 ft.

C   16 ft.

D   32 ft.

Know It All Approach

First, read the question carefully and make note of the words and information that you need in order to find the correct answer. Whenever a question has a diagram (e.g., a picture, graph, table, or chart), you will probably need the information in the diagram to figure out the correct answer. In this case, take note of the scale (one box = 0.25 inches), the length of the killer whale (4 inches), and the length of the ceiling tile (0.25 inches). Also, notice that the question reads "actual ceiling tile measures 24 inches on each side," and "what is the actual length (in feet) of the killer whale?"

Now that you have all the information you need, answer the question. The easiest way to figure out the answer is to set up a proportion.

$$\frac{\text{scale length of ceiling tile}}{\text{actual length of ceiling tile}} = \frac{\text{scale length of killer whale}}{\text{actual length of killer whale}}$$

$$\frac{0.25 \text{ in.}}{24 \text{ in.}} = \frac{4 \text{ in.}}{x}$$

Next, cross multiply.

$$0.25x = 24 \times 4$$

$$0.25x = 96$$

Now, divide both sides by 0.25.

$$x = 384 \text{ in.}$$

Is this your final answer? No, it is not! You have to convert the answer to feet.

$$384 \text{ in.} \div 12 = 32 \text{ ft.}$$

Now, check your answer. A good way to do this is to reverse each process. Multiply 32 by 12, and you get 384 inches. Next, set up the same proportion, this time substituting 384 inches for x. Finally, make sure that the two ratios are equal.

$$\frac{0.25}{24} = \frac{4}{384}$$

$$\frac{1}{96} = \frac{1}{96}$$

If you were not able to calculate the answer, you could increase your chances of guessing the correct answer by reading all of the answer choices and eliminating those choices you know are incorrect. The ceiling tile is 2 feet (24 in.) long, so answer choice (A) makes no sense. Similarly, answer choice (B) is too small. Answer choices (C) and (D) both seem reasonable. However, use some common sense. If the image of the killer whale is 16 boxes long, and each box represents 2 feet (24 inches), then the actual killer whale must be 16 × 2, or 32 feet long. You can eliminate (C), so choice (D) is the correct answer.

Directions: Read the passage below and answer the questions on the next page.

The Hole in the Sky

 Over the last twenty years, the ozone layer covering Earth has decreased, particularly over Antarctica. The ozone layer forms a natural screen that absorbs most of the UV (ultraviolet) radiation from sunlight *before* it reaches Earth's surface. Too much UV radiation causes skin cancer. Man-made chemicals (CFCs) used in refrigerants and aerosol sprays have caused most of the damage. The ozone hole is a damaged area of ozone layer over Antarctica.

In New Zealand, a country near Antarctica, it is estimated that ozone losses since 1980 have caused UV radiation to increase by 10 to 12 percent. New Zealand has the highest rate of skin cancer in the world, with almost 50,000 new cases every year. All New Zealanders, especially children of school age, have been advised to avoid the sun between 11:00 A.M. and 4:00 P.M., particularly in summer. If they do go outside, they should apply sunscreen and wear sun-protective clothing, such as hats and long-sleeved shirts. The good news is that, thanks to recent laws banning the use of CFCs, the ozone hole is shrinking. In about fifty years, ozone levels should be back to normal.

1. At a school in New Zealand, the Parents Teachers Association (PTA) is raising money to plant trees to provide more shade from the sun. They would like to plant a total of twenty trees such that the ratio of maple trees to oak trees is 3 to 2. How many trees will be oak trees?

    A 4
    B 6
    C 8
    D 12

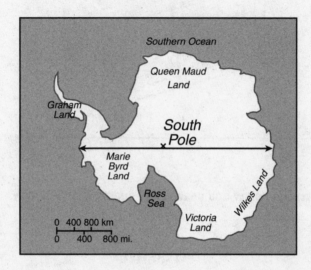

2. Above is a map of Antarctica. The line with an arrow at each end cuts straight across the continent and passes through the South Pole. According to the scale of kilometers located at the bottom of the map, what is the approximate width of Antarctica along this line?

    _____

3. The actual depth, front to back, of the scale building shown below is 24 meters. Find the actual width, side to side, using the scale drawing.

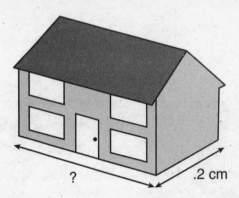

? .2 cm

Directions: Read the passage below and answer the questions that follow.

Cheater!

In the 84th Boston Marathon on April 21, 1980, amateur runner Rosie Ruiz came from out of nowhere to win the women's race. She had achieved the third-fastest time ever recorded for a female runner. Observers noticed that she looked remarkably sweat-free and relaxed as she crossed the finish line. However, twenty-four hours later, questions arose when it was found that Ruiz did not appear in race videotapes until near the end of the race. Monitors at the various race checkpoints had not seen her, nor had any of the other runners. Eventually, some members of the crowd came forward, saying they had seen Ruiz jump into the race during its final half-mile.

The accepted theory came to be that Ruiz had hopped a subway for much of the race, ran the final half-mile, and then "proudly" accepted the winner's medal. A week later, Ruiz, who never admitted to cheating, was officially disqualified from the race, and Jacqueline Gareau was declared the true winner.

4. The ratio of male runners to female runners in a marathon was 5 to 2. If the total number of runners was 21,000, how many of them were female?

   A 4,200

   B 6,000

   C 8,400

   D 15,000

5. Alex is building a large glider out of balsa wood. She made a scale drawing of the airplane that has a total length, from nose to tail, of 8.5 inches, with each wing 5 inches long.

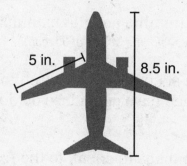

5 in.   8.5 in.

If each wing on the full-size glider will be two feet long, what will be the length from nose to tail of the full-size glider **in feet**? Use the scale drawing to help you figure out your answer.

   A 2.6 ft.

   B 3.0 ft.

   C 3.4 ft.

   D 4.2 ft.

Subject Review

In Chapter 5, you learned that a ratio is a comparison of two numbers by division. In addition, you found out how to solve for an unknown quantity in a ratio by setting up a proportion. A proportion is an equation that shows that two ratios are equivalent. Lastly, you were told that scale is the ratio of a given length on a drawing or model to its corresponding length in reality. You learned that measurements on a scale model or drawing are proportional to measurements on the actual object.

You also learned some interesting facts.

What is the fastest-swimming aquatic mammal?
The orca, or killer whale, can swim at a speed of 30 mph.

What country requires school-aged children to wear hats and long-sleeved shirts when they have gym class outdoors?
New Zealand

How did Rosie Ruiz manage to sneak to the front of the line in the 1980 Boston Marathon?
She took the subway!

Chapter 6: Exponents and Square Roots

How many skydivers participated in the world's largest freefall display?

What is the most distant object in the universe?

How tall is the giant ketchup bottle in Collinsville, Illinois?

Exponents

An **exponent** is a small number placed next to and above another number (called the **base**) to show how many times that number is to be multiplied by itself. You can use exponents to show repeated multiplication. In the example below, "2" is the base and "5" is the exponent. The expression "2^5" is read as "two to the fifth power."

$$2^5 = 2 \cdot 2 \cdot 2 \cdot 2 \cdot 2 = 32$$

It is sometimes easier to write large numbers if you use exponents. For example, one billion is usually written as 1,000,000,000. Using exponents, you would write one billion as 10^9. Exponents may be positive, negative, or zero. Furthermore, variables as well as numbers may have exponents.

Computing with Exponents

There are certain rules you must follow when doing computations with exponents.

1. To multiply numbers or variables with the same base, add the exponents.

    $2^3 \cdot 2^4 = 2^7$ $\qquad$ $a^2 \cdot a^3 = a^5$

2. To divide numbers or variables with the same base, subtract the exponents.

    $3^5 \div 3^2 = 3^3$ $\qquad$ $\dfrac{x^7}{x^3} = x^4$

3. To find the power of a power, multiply the exponents.

    $(2^3)^4 = 2^{12}$ $\qquad$ $(y^2)^3 = y^6$

4. To simplify a number or variable with a negative exponent, invert the number and make the exponent positive.

$$4^{-2} = \frac{1}{4^2} \qquad b^{-3} = \frac{1}{b^3}$$

5. Any nonzero number or variable with an exponent of zero is equal to one.

$$3^0 = 1 \qquad n^0 = 1$$

Scientific Notation

Scientists often have to work with very large or very small numbers. For example, the speed of light is 300,000,000 meters per second, and the mass of a hydrogen atom is

0.000000000000000000000001672 grams. To make numbers such as these easier to work with, scientists rewrite them using exponents in what is referred to as **scientific notation.**

To express a number in scientific notation, you write it as the product of a number between one and ten and a power of ten. If the number is very large, you move the decimal point to the left until you have a number between one and ten, and then multiply it by the power of ten equal to the number of places you moved the decimal point.

$$\frac{300,000,000 \text{ m}}{\text{sec.}} = 3.0 \times \frac{10^8 \text{ m}}{\text{sec.}}$$

If the number is very small, you move the decimal point to the right until you have a number between one and ten, and then multiply it by the negative power of ten equal to the number of places you moved the decimal point.

$$0.000000000000000000000001672 \text{ grams} = 1.672 \times 10^{-24} \text{ grams}$$

To convert from scientific notation to standard notation, you do the reverse. Get rid of the exponent, and move the decimal point either to the right (for a positive exponent) or to the left (for a negative exponent) the number of places equal to the absolute value of the exponent.

$$2.35 \times 10^4 = 23,500 \qquad 1.8 \times 10^{-3} = 0.0018$$

Squares and Square Roots

The **square** of a number is the number multiplied by itself. For example, the square of 9 is 81 ($9 \bullet 9$ or $9^2 = 81$). The square of -5 is 25 ($(-5) \bullet (-5)$ or $(-5)^2 = 25$).

The **square root** of a given number is a number that, when multiplied by itself, equals the given number. The square root of 49 is 7 ($7 \bullet 7$ or $7^2 = 49$). Technically, the square root of 49 is $+/-7$, since $(-7) \bullet (-7)$ or $(-7)^2$ is also equal to 49. The symbol for square root is "$\sqrt{}$" (for example, $\sqrt{16} = +/-4$).

Some numbers are perfect squares. The square root of a **perfect square** is an integer, or a whole number. Thus, 1, 4, 9, 16, 25, etc., are perfect squares. Most numbers are not perfect squares.

You can estimate the square root of a number. For example, what is the square root of 70? You may know that $\sqrt{64} = 8$, and $\sqrt{81} = 9$. You can estimate that $\sqrt{70}$ is somewhere between 8 and 9. Moreover, it is probably closer to 8 because 70 is closer to 64 than it is to 81. Thus, you can estimate that $\sqrt{70}$ is about 8.3 or 8.4.

Directions: Read the passage below and answer the question that follows.

Extreme Team Skydiving

The World Team is a large group of skydivers who get together on a regular basis to break world records in skydiving. The team includes members from thirty-five different countries. They met in Slovakia in 1994, and again in Russia in 1996. On December 16, 1999, the team completed a freefall formation of 282 skydivers in the skies above Thailand. The skydivers held the link for 7.11 seconds. In December of 2002, skydivers from twenty-five different countries broke the Team World record, assembling a human formation of three hundred parachutists. How long will this record last? The World Team is already planning a larger record-breaking jump and who will finally hold the record for largest skydiving formation is anybody's guess.

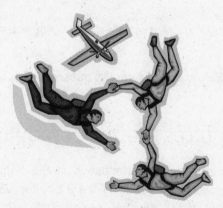

▶ The formula for the distance a free-falling body falls in a given time is $d = \frac{1}{2}at^2$, in which **d** is the distance, **a** is the acceleration due to gravity (32 ft./sec./sec.), and **t** is the time. What distance does a skydiver fall in the first three seconds of free fall?

Use the **Know It All Approach** to find your answer. First, read the question carefully and make note of the important words and information. The formula $d = \frac{1}{2}at^2$ and the numbers "32 ft./sec./sec." and "three seconds" are important in order to answer the question. You need to find distance.

After you've figured out what you need to find, and you've found all the necessary information, calculate the answer. Write the formula, substitute the numbers given in the question, and then solve for *d*.

$$d = \frac{1}{2}at^2$$

$$d = \frac{1}{2} \bullet 32 \bullet 3^2$$

$$d = 16 \bullet 9$$

$$d = 144 \text{ ft.}$$

Before you write down your final answer, check your work. The easiest way to do this is to substitute your answer for *d* in the formula, and then check to see if both sides of the equation are equal.

$$d = \frac{1}{2}at^2$$

$$144 = \frac{1}{2} \bullet 32 \bullet 3^2$$

$$144 = 16 \bullet 9$$

$$144 = 144$$

Once you've checked your answer and you're sure that it is correct, write the answer clearly in the space provided.

Directions: Read the passage below and answer the questions that follow.

Pictures from Deep Space

In April 2000, astronomers participating in the Subaru Deep Field (SDF) project began collecting data using the 8.2-meter (26.9-ft.) optical-infrared telescope located at Mauna Kea, Hawaii. The astronomers discovered a galaxy with a red shift of 6.58. Red shift is one of the factors astronomers use to gauge the enormous distances involved in studying space. As the light given off by distant galaxies makes its long journey across the universe, it shifts to longer or "redder" wavelengths. This phenomenon is called red shift. A galaxy's red shift indicates how far away it is — the greater the red shift, the greater the distance. Astronomers calculated that this large red shift corresponds to a distance of around 12.8 billion light years — making this newly discovered galaxy the farthest object yet located in the universe. They also state that the galaxy gives them a picture of what the universe was like when it was just 900 million years old. The SDF team expects to find many more distant galaxies through continued observations.

1. What is 12,800,000,000 (12.8 billion) written in scientific notation?

    A 1.28×10^9
    B 1.28×10^{10}
    C 128×10^9
    D 12.8×10^{10}

2. An astronomer compared the diameters of the four stars listed in the chart below.

Star	Diameter (km)
1	1.39×10^6
2	4.18×10^8
3	1.50×10^9
4	3.34×10^7

Which of these stars has the largest diameter?

    A   Star 1
    B   Star 2
    C   Star 3
    D   Star 4

3. $\dfrac{10^5}{10^{-2}} =$

    A 10^{-10}
    B 10^{-3}
    C 10^3
    D 10^7

Saving the World's Largest Ketchup Bottle

The world's largest ketchup bottle may well be the 170-foot tall, ketchup bottle-shaped water tower located next to Route 159, just south of Collinsville, Illinois. It was built in 1949 by the W. E. Caldwell Company for Brooks Foods, bottlers of Brooks ketchup. The company sold the plant in the early 1970s, and the new owner had no use for the water tower. It was allowed to deteriorate for more than twenty years, causing the giant bottle to peel and rust. Eventually, it was scheduled for demolition. Over a two-year period, the Ketchup Bottle Preservation Group sold nearly six thousand shirts and petitioned for pledges, not only in Collinsville, but across the entire USA. More than $70,000 was raised to restore this roadside landmark attraction to its original appearance. A parade and lighting ceremony were held in June 1995. Recognized the world over as an excellent example of twentieth-century roadside Americana, the "World's Largest Ketchup Bottle" was named to the National Register of Historic Places in August, 2002.

4. The Pythagorean theorem can be used to determine the length of an unknown side of a right triangle. Seema used it to figure out the height of a water tower. She calculated that the water tower was $\sqrt{22{,}500}$ feet tall. Which of the following should she use to report the height of the water tower?

   A $\pm$ 150 ft.

   B 150 ft.

   C -150 ft.

   D 150^2 ft.

5. The square root of 270 is a number between

   A 13 and 14

   B 14 and 15

   C 15 and 16

   D 16 and 17

6. $\sqrt{9x^2} =$

   A 3

   B $3x$

   C $9x$

   D $3x^2$

7. $5^3 \times 5^2 =$

   A 5^5

   B 5^6

   C 25^5

   D 25^6

Subject Review

In this chapter, you learned that an exponent is a number placed next to and above another number to show how many times that number is to be multiplied by itself.

You also were given rules to follow when computing with exponents. You learned how to express very large or very small numbers using scientific notation.

You also learned that the square of a number is the number multiplied by itself, and that the square root of a given number is another number that, when multiplied by itself, equals the given number.

Finally, you've learned how to answer the three questions at the beginning of the chapter.

How many skydivers participated in the world's largest freefall display?

300 skydivers jumped at one time together in 2002, but the World Team hopes to break that record in 2004.

What is the most distant object in the universe?

A galaxy was found to be 12.8 billion light years from Earth.

How tall is the giant ketchup bottle in Collinsville, Illinois?

It is 170 feet tall.

Directions: Read the passage below and answer the questions that follow.

The Blackbird

The SR-71 "Blackbird" is a military airplane produced by Lockheed-Martin Aeronautics Company. It was developed in response to a need for improved aerial surveillance and photography. After the infamous 1960 U-2 incident, in which a spy plane piloted by Francis Gary Powers was shot down over the former Soviet Union, it was decided that future reconnaissance aircraft should be capable of flying at speeds above Mach 3 (2,150 miles per hour) and higher than 85,000 feet. Such aircraft would be able to outrun or outdistance any jet interceptor or surface-to-air missile. The "Blackbird" is the world's fastest and highest-flying aircraft. From an altitude of 80,000 feet, it is capable of surveying 100,000 square miles of Earth every hour.

In July 1976, the SR-71 set two world records: an absolute speed record of 2,193 miles per hour and an absolute altitude record of 85,069 feet. In 1990, the "Blackbird" was retired because the operating costs were very high. However, it was briefly brought back into service in 1995. All "Blackbirds" were permanently retired in 1998 after the U.S. Department of Defense decided that satellites orbiting Earth in outer space could perform all necessary surveillance.

1. The SR-71 Blackbird flew 4,500 miles at an average speed of 2,000 miles per hour. How long did it take to cover this distance?

   A $\frac{1}{2}$ hr.

   B $1\frac{1}{2}$ hr.

   C $2\frac{1}{4}$ hr.

   D 9 hr.

2. If you multiply 4 by an integer less than -2, which of the following will be the result?

   A an integer greater than -2
   B an integer less than -8
   C an integer between -8 and 8
   D an integer greater than 8

3. If $-|n| = -5$, what is the value of n?

   A -5 or 0
   B 0 or 5
   C -5 or 5
   D -25 or 25

4. Simplify the following: $-|-7 + 2| - |-3|$.

   A -8
   B -2
   C 2
   D 8

5. Simplify: $x + 2y - (x - y)$.

   A y

   B $2x + y$

   C $3y$

   D $2x + 3y$

6. Simplify the following: $3^2 \bullet 4 \div 6 + (10 - 2^3)$

   A -2

   B 0

   C 4

   D 8

7. The eight members of the cheerleading squad are going out to lunch. Each cheerleader will get a sandwich ($2.25), a salad ($1.25), and a juice ($1.50). What will be the total cost of the meal for all the cheerleaders? **Show all work**.

_____

Directions: Read the passage below and answer the questions that follow.

Space Tourist

Space tourist Dennis Tito paid $20 million to Russia for a six-day visit to the International Space Station in 2001. Tito was transported to the station aboard a Russian Soyuz spacecraft. He had to sign agreements absolving all participating countries from blame in the event he was injured or killed. "The big surprise was that, despite a small bout of space sickness the first day, the rest of the trip was a major high," Tito explained. He said the experience was very relaxing. "I could have stayed up there for months. I was not at all bored."

8. Recently, a famous music entertainer expressed a desire to go on a space voyage. NASA handed him a 792-page training manual. Over the first 5 days, he read 120 pages. If he continues reading at the same rate, how many more days will it take him to finish the manual?

A   12
B   24
C   28
D   33

9. List the following numbers in order from smallest to largest:

$\frac{5}{8}$, 5.8, 58%

A $\frac{5}{8}$, 5.8, 58%

B   5.8, 58%, $\frac{5}{8}$

C $\frac{5}{8}$, 58%, 5.8

D   58%, $\frac{5}{8}$, 5.8

10. If Keyshawn goes to school 171 of 180 days, what is his attendance percentage?

A   5%
B   17%
C   91%
D   95%

11. Last month, a dress was selling for $55.00. This month, the same dress costs $44.00. What is the percent decrease in price?

    A   20%

    B   25%

    C   75%

    D   80%

12. $\frac{15}{18} - \left(\frac{2}{3} - \frac{1}{6}\right)$

    A   0

    B $\frac{1}{6}$

    C $\frac{1}{3}$

    D $\frac{4}{3}$

13. The ratio of girls to boys on the volleyball team is 5 to 3. If the team has 24 players, how many of them are boys?

    A   3

    B   6

    C   9

    D   15

14. The model car shown below is $5\frac{1}{2}$ inches long and was made using a scale of 1:30. How long (in **feet**) is the actual car?

   A $13\frac{3}{4}$ ft.

   B $17\frac{1}{4}$ ft.

   C   18 ft.

   D   162 ft.

15. $\dfrac{10^{-3}}{10^{-5}} =$

   A 10^{-8}

   B 10^{-2}

   C 10^{2}

   D 10^{15}

Chapter 7: Variables and Algebraic Expressions

How many people participated in the world's largest snowball fight?

What is the highest fall without a parachute in which the jumper survived?

Who was the first gymnast to score a perfect 10 at the Olympics?

Variables

A **variable** is a symbol, usually a letter, used to represent a number in mathematical expressions or sentences. For example, in Chapter 3, you used the formula "$d = s \times t$," or distance equals rate of speed × time. The letters d, s, and t are variables. If you know the value of two of these variables, you can figure out the value of the third (unknown) variable. Try the sample question below.

▶ Migdalia drove with three friends from her house to the ski resort. The trip took $3\frac{1}{2}$ hours driving at an average speed of 60 miles per hour. How far away is the ski resort from Migdalia's house?

The question asks you to find the distance to the ski resort. Let d represent the unknown distance. Since the question gives you the values of the other variables (s and t), just plug them into the formula and solve for d.

$$d = s \times t$$

$$d = 60 \cdot 3\frac{1}{2}$$

$$d = 210 \text{ miles}$$

Evaluate means "find the value of." To evaluate a mathematical expression, replace each variable with the corresponding value given in the question.

Algebraic Expressions

An **algebraic expression** is a combination of variables, numbers, and at least one mathematical operation. For example, $2n + 5$, $y - 3x$, and $a^2 + b^2$ are all algebraic expressions. Solving word problems in algebra depends on your ability to represent missing or unknown quantities based on their description in the question. For example, consider the sample question below.

▶ A vendor has x hot dogs. He sells thirty-two hot dogs, and then picks up a new supply of $3y$ hot dogs at the butcher. Which expression represents the number of hot dogs he now has?

   A $x - 32 + 3y$
   B $x - 32 - 3y$
   C $x + 32 + 3y$
   D $x + 32 - 3y$

You need an algebraic expression that represents the scenario described in the question. If the vendor had x hot dogs and he sold thirty-two, then he had "$x - 32$" hot dogs left. Then he picked up $3y$ new hot dogs. So, add the new hot dogs to the ones he had left. You get the expression "$x - 32 + 3y$".

The expression "$3y$" is an example of a monomial. A **monomial** is a number, a variable, or a product of a number and one or more variables. A monomial does not contain addition or subtraction. Thus, $9n^2$, 5, $2ab$, and $4x^3y^5$ are all monomials. Try this sample question about monomials.

▶ What is the product of $(2x^2y)(5x^3y^4)$?

   A $7x^4y^6$
   B $7x^3y^6$
   C $10x^3y^6$
   D $10x^5y^5$

Because monomials have no $+$ or $-$ signs, all the numbers and variables are "factors." Therefore, all you need to do is multiply all the factors to find the answer.

$$(2x^2y)(5x^3y^4) = 2 \cdot x^2 \cdot y \cdot 5 \cdot x^3 \cdot y^4$$

Now, rearrange factors, grouping numbers with numbers and like variables with like variables.

$$(2x^2y)(5x^3y^4) = 2 \cdot 5 \cdot x^2 \cdot x^3 \cdot y \cdot y^4$$

The final step is to multiply, making sure you follow the rules of exponents.

$$(2x^2y)(5x^3y^4) = 10 \cdot x^5 \cdot y^5 = 10x^5y^5$$

Therefore, answer choice (D) is correct.

The expression "$x - 32 + 3y$" is an example of a polynomial. A **polynomial** is the sum or difference of two or more monomials. Thus, "$10x + 3y$" and "$a^2 + 12a + 20$" are polynomials. Multiplying polynomials is a lot more complicated than multiplying monomials. Look at the sample question below.

▶ Simplify: $(x + 7)(x + 2)$

   A $2x + 9$
   B $x^2 + 14$
   C $x^2 + 9x + 14$
   D $x^2 - 5x + 14$

Be careful! You cannot just multiply, for example, $x \cdot 7 \cdot x \cdot 2$. These are not factors because they have $+$ signs between them. The factors are the polynomials "$(x + 7)$" and "$(x + 2)$." In order to multiply them, you have to use the FOIL method. **FOIL** stands for "First-Outside-Inside-Last."

Multiply the two **First** parts: $x \cdot x = x^2$

Multiply the two **Outside** parts: $x \cdot 2 = 2x$

Multiply the two **Inside** parts: $7 \cdot x = 7x$

Multiply the two **Last** parts: $7 \cdot 2 = 14$

The final step is to add together like terms: $x^2 + 2x + 7x + 14 = x^2 + 9x + 14$. Therefore, answer choice (C) is correct.

Factoring

Situations will come up in which you know a product and you need to know the factors that resulted in that product. Finding the factors is called **factoring**. Sometimes, factoring involves simply finding the greatest common factor (GCF). Try this sample question.

▶ What is the GCF of $12ab^2$ and $9a^2b$

   A $3ab$
   B $4ab^2$
   C $3a^2b$
   D $4a^2b^2$

Find the GCF for the numbers and each variable, and then combine them.

What is the GCF of 12 and 9?   The greatest factor common to 12 and 9 is 3.

What is the GCF of a and a^2?   The greatest factor common to a and a^2 is a.

What is the GCF of b and b^2?   The greatest factor common to b^2 and b is b.

Thus, the GCF of $12ab^2$ and $9a^2b$ is $3ab$, and answer choice (A) is correct.

Another factoring trick is the difference between two perfect squares. The rule is: $a^2 - b^2 = (a + b)(a - b)$. Try the sample question below.

▶   Factor: $4x^2 - 25$

   A $(2x - 5)$
   B $4(x + 1)(x - 5)$
   C $(2x + 5)(2x - 5)$
   D $(2x - 5)(2x - 5)$

The question gives you $4x^2 - 25$, which is the difference between two perfect squares ($4x^2$ and 25).
First, find the square root of each perfect square: $\sqrt{4x^2} = 2x$, $\sqrt{25} = 5$. Then, follow the rule:
$4x^2 - 25 = (2x + 5)(2x - 5)$. Hence, answer choice (C) is correct.

Factoring a trinomial (a polynomial with three terms) is a bit trickier. Study the sample question below.

▶   Factor: $2x^2 + x - 10$

   A $(x + 5)(x - 2)$
   B $(2x + 5)(x - 2)$
   C $(2x - 5)(x + 2)$
   D $(2x + 1)(x - 10)$

The process involved here is the reverse of FOIL. To get the first term of $2x^2$, the only possible factors are $2x$ and x. So you start with: $(2x)(x)$. To get the last term of -10, there are many possible factors. You need to find the combination that produces the middle term of $+ x$.

$+ 10$ and $- 1$	$(2x + 10)(x - 1)$	$10x - 2x = 8x$	Wrong.
$- 10$ and $+ 1$	$(2x - 10)(x + 1)$	$-10x + 2x = -8x$	Wrong.
$+ 5$ and $- 2$	$(2x + 5)(x - 2)$	$5x - 4x = x$	Right!
$- 5$ and $+ 2$	$(2x - 5)(x + 2)$	$-5x + 4x = -x$	Wrong.

The factors are $(2x + 5)(x - 2)$, and answer choice (B) is correct.

Multiplying a Monomial and a Polynomial

The last thing you need to know about algebraic expressions is how to multiply a monomial and a polynomial. Check out the sample question below.

▶ Simplify: $3ab(4b + 3a)$

A $\quad 12a + 9b$
B $\quad 12b^2 + 9a^2$
C $\quad 12ab^2 + 3a^2b$
D $\quad 12ab^2 + 9a^2b$

If you recall, the distributive property states that: $a(b + c) = ab + ac$. So all you have to do is distribute the factors over the terms inside the parentheses.

$$3ab(4b + 3a) = 3ab(4b) + 3ab(3a) = 12ab^2 + 9a^2b$$

Therefore, answer choice (D) is correct. If the answer looks familiar, it's because you did the reverse procedure earlier when you found the GCF. Finding the GCF is the reverse of applying the distributive property.

Directions: Read the passage below and answer the question that follows.

Snowball War

 The largest snowball fight in history occurred on January 18, 2003, at Triel in the ski resort town of Obersaxen-Mundaun in Graubunden, Switzerland. The 2,473 participants were broken up into two teams—the red team with 1,162 members, and the blue team with 1,311 members. Eight security marshals supervised the event, keeping each team behind a barrier of yellow tape. A distance of 20 meters (65.5 feet) between the two teams was maintained throughout the contest. The two teams pummeled each other with snowballs for ten minutes. It was all done in good clean fun, and there were no reported injuries. Reporters from the local and European press witnessed the event.

▶ If each member of the red team (1,162 members) threw x snowballs, and each member of the blue team (1,311 members) threw y snowballs, which of the following expressions represents the total number of snowballs thrown?

A $\quad x + y$
B $\quad x - y$
C $\quad 1,162x + 1,311y$
D $\quad 1,162x - 1,311y$

Know It All Approach

Find the right answer using the **Know It All Approach.** First, read the question carefully and make note of the words and information that you need in order to find the correct answer. The important information would be x snowballs for each member of the red team and y snowballs for each member of the blue team. In addition, you should make note of the numbers "red: 1,162" and "blue: 1,311."

Now, answer the question. If each red team member threw x snowballs, then the entire red team threw $1,162x$ snowballs. Similarly, if each blue team member threw y snowballs, then the entire blue team threw $1,311y$ snowballs. The total number of snowballs thrown would be the sum of the two teams' totals.

$$1,162x + 1,311y$$

Next, check your answer. This is not a computation, so there is no mathematical way to check your answer. All you can do is look it over carefully, walk through the steps you took to solve the answer, and decide if your solution makes sense.

Remember to read all the answer choices, and eliminate those that you *know* are incorrect. Choice (A) would equal the sum of the number of snowballs thrown by one member of each team. This would be a relatively small number, and thus could not possibly represent the total number of snowballs thrown. Thus, you can eliminate (A). Choice (B), a subtraction, would be even smaller, so you can eliminate (B). Choice (C) is the sum of the number of snowballs thrown by the red team and the blue team, so it is most likely correct. Choice (D) is a subtraction, and you don't get a **total** by subtracting. Hence, you can eliminate (D). As suspected, answer choice (C) is correct.

Directions: Read the passage below and answer the questions that follow.

Any Landing You Can Walk Away From . . .

Vesna Vulovic, a flight attendant from Yugoslavia, survived a fall from 33,330 feet (10,160 meters) when the DC-9 airplane she was traveling in crashed over the Czech Republic. A former nurse, Bruno Henke, spotted Vesna's legs sticking out of the plane's cracked fuselage. Bruno cleared her airway, and then rushed her to a hospital. Vesna awoke from a coma three days later. She suffered no permanent injuries.

A Russian pilot survived a fall of 21,980 feet (6,700 meters) when he jumped from a damaged plane **without a parachute!** He landed on the side of a snow-covered mountain and slid to the bottom, breaking his pelvis and injuring his spine. Fortunately, he made a full recovery.

1. Consider the Russian pilot who jumped from a damaged plane without a parachute. You can use the following expression to calculate the time it took him to fall to the earth: $d = \frac{1}{2} at^2$, where d = the distance he fell, a = the acceleration due to gravity (32 ft./sec./sec.), and t = the time it took him to fall. If he fell a distance of 21,980 feet, how much time did it take him to reach the ground? Round your answer to the nearest second.

   _____

2. If $n = -|-5|$, then $-n =$

   A   5

   B   −5

   C $\frac{1}{5}$

   D $-\frac{1}{5}$

3. After she arrived at the airport, Ms. Ellsworth needed to rent a car. She knew that the cost C of renting a car varies according to the formula: $C = \$45d + \$0.15m$, where d is the numbers of days rented and m is the number of miles driven. If she rented a car for 3 days and drove 280 miles, what was the cost of the rental?

   A $45.15

   B $97.00

   C $135.15

   D $177.00

4. Simplify: $(x + 3)(2x − 1)$.

   A $2x^2 − 3$

   B $2x^2 − x − 3$

   C $2x^2 + 5x − 3$

   D $2x^2 − 5x − 3$

5. If $z = -2xy$, $z = 24$, and $x = 4$, what is the value of y?

   A   −3

   B   3

   C   4

   D   12

Perfect 10

Born in 1961 in Romania, Nadia Comaneci was accepted into Gymnastics High School at age eight. Her schedule called for five hours in the classroom and four hours in the gym every day. She arrived at the 1976 Olympics in Montreal, Canada with little hope of taking home a gold medal. Then she performed her routine on the uneven parallel bars. She scored a perfect 10! It was the first time in history that any gymnast received a perfect score. She would score six more 10s before the end of the Olympics. She left Montreal with three gold medals, one silver, and one bronze.

In the 1980 Olympics in Moscow, Nadia Comaneci nabbed two more gold medals and two more silver medals.

The young gymnast retired in 1984 and later became an international judge and a coach. In 1989, she defected to the United States and now lives in Oklahoma. She married Bart Connor, who is also an Olympic gold medalist.

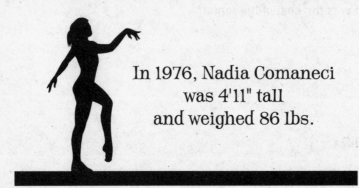

In 1976, Nadia Comaneci
was 4'11" tall
and weighed 86 lbs.

6. Which of the following expressions represents the product of Nadia's score times the product of a and b?

   A $\dfrac{ab}{10}$

   B $10ab$

   C $\dfrac{10}{ab}$

   D $\dfrac{b}{10a}$

7. Factor completely: $3x^3 + 9x^2 - 12$

   A $3x(x + 2)(x + 2)$
   B $3x(x + 2)(x - 2)$
   C $3x(x + 4)(x - 1)$
   D $3x(x - 4)(x + 1)$

8. Which of these expressions is equivalent to $3(z + 4) - 2(z + 1) = 13$?

   A $3z + 12 - 2z - 1 = 13$
   B $3z + 4 - 2z + 1 = 13$
   C $3z + 4 - 2z + 2 = 13$
   D $3z + 12 - 2z - 2 = 13$

9. The product of $(5a^2b)(3a3b^4)$ equals

   A $8a^3b^6$
   B $8a^2b^6$
   C $15a^2b^6$
   D $45a^2b^5$

10. Factor completely: $7x^2 - 28$

   A $7(x + 2)(x - 2)$
   B $7(x - 2)(x - 2)$
   C $7(x - 4)^2$
   D $7(x - 4)$

Subject Review

In Chapter 7, you learned that a variable is a symbol used to represent a number in mathematical expressions. You also learned that an algebraic expression is a combination of variables, numbers, and operational signs. Lastly, you learned how to compute with monomials and polynomials.

You also learned a couple of cool facts.

How many people participated in the world's largest snowball fight?
2,473 people threw snowballs at each other in a giant snowball fight in 2000.

What is the highest fall without a parachute in which the jumper survived?
Russian pilot I.M. Chisou fell 21,980 feet (6,700 meters) and survived. A snowy landing helped cushion the fall.

Who was the first gymnast to score a perfect 10 at the Olympics?
Nadia Comaneci scored seven perfect scores in her professional gymnastics career. In 1976, she was the first person to do so in the Olympic games.

Chapter 8: Patterns and Functions

What was René Descartes watching when he came up with the idea of the x- and y-coordinate graph?

Who was the first seeing-eye™ dog?

What gets tossed around during the traditional festival of La Tomatina?

Number Patterns

A **pattern** is a series of numbers or figures that follows a rule. Look at the series of numbers below.

0, 3, 6, 9, 12, 15, 18, 21, 24, 27, ?

What should be the next number in the series? To find the answer, you need to figure out what rule the series of numbers follows. For this particular pattern, the rule is to add 3 to each number to get the next number. Therefore, the next number in the series should be 27 + 3, or 30.

The rules for some number patterns are not as basic as the one above. Check out the next series of numbers.

1, 1, 2, 3, 5, 8, 13, 21, 34, ?

What should be the next number in this series? You may find it a little more challenging to identify the rule for this series of numbers. Look at the two numbers preceding each number to try to find the rule. The rule is to add the two preceding numbers to get the next number. So consider the numbers in the pattern again, now that you know the rule.

1, 1 (1 + 0), 2 (1 + 1), 3 (1 + 2), 5 (2 + 3), 8 (3 + 5) . . .

Therefore, the next number in the series should be 55 (21 + 34).

Geometric Patterns

Not all patterns involve a series of numbers. Geometric patterns consist of a series of geometric figures. Try the sample question on the next page. Study the series of figures, and see if you can figure out the pattern.

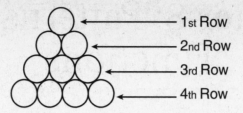

1st Row
2nd Row
3rd Row
4th Row

▶ If the pattern continues, how many circles should be in the seventh row?

A 6
B 7
C 8
D 10

Count the number of circles in each row. The first row has one circle, the second row has two circles, the third row has three circles, and the fourth row has four circles. So the seventh row must have seven circles. Answer choice (B) is correct. Now, try to identify the rule in the next geometric pattern.

▶ Shown below is a mosaic Todd created. He surrounded white tiles with shaded tiles.

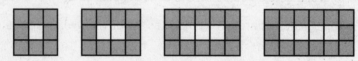

If the pattern continues, how many shaded tiles will Todd need when there are eight white tiles in the center?

A 16
B 18
C 22
D 24

How do you find the answer to this one? Once again, you need to figure out the rule the pattern follows. If the pattern is not obvious, a good way to figure it out is to make a table.

Number of white tiles	1	2	3	4
Number of shaded tiles	8	10	12	14

Notice that the number of shaded tiles, or *s*, is equal to twice the number of white tiles, or *w*, plus six, or $s = 2w + 6$. You can use this rule to figure out the number of shaded tiles needed when there are eight white tiles. $S = (2 \cdot 8) + 6 = 22$. Therefore, answer choice (C) is correct.

Functions

The pattern of tiles shown in the question above is not only a pattern—it's a function! A **function** is a relationship specified by a rule that pairs up each input with a corresponding output. In a function, a rule dictates that an operation be done on one number to produce another number. Because the **output** depends on the **input**, you can say that the output "is a function of" the input. Often, the input is shown as *x* and the output is shown as the function of *x*, or $f(x)$.

Look at the table you made for the last question. You could rewrite the table in the form of a function.

Input: *x*	Rule: 2*x* + 6	Output: *f(x)*
1	$2 \cdot 1 + 6$	8
2	$2 \cdot 2 + 6$	10
3	$2 \cdot 3 + 6$	12
4	$2 \cdot 4 + 6$	14

You can write functions as equations. For example, you can write $f(x) = 2x + 6$. Now, look back at the previous question from page 80. How many shaded tiles are needed when there are eight white tiles in the center? To find the answer, just input the number of white tiles into the equation: $f(8) = 2 \cdot 8 + 6 = 22$.

You arrived at the equation by examining the values in the table. Next, try doing the reverse. You will be given an equation of a function and then will be asked to choose the corresponding table of values.

▶ Which of the following tables represents the function $f(x) = 3x - 1$?

x	f(x)
1	4
2	7
3	10
4	14

A

x	f(x)
1	2
2	5
3	8
4	11

B

x	f(x)
1	2
2	3
3	4
4	5

C

x	f(x)
1	5
2	7
3	9
4	11

D

You can find the answer to this question by simply plugging in the values for *x*.

$f(x) = 3x - 1$

$f(1) = 3x - 1 = 3 \cdot 1 - 1 = 2$

$f(2) = 3x - 1 = 3 \cdot 2 - 1 = 5$

$f(3) = 3x - 1 = 3 \cdot 3 - 1 = 8$

$f(4) = 3x - 1 = 3 \cdot 4 - 1 = 11$

You can see that the values for $f(x)$ when $x = 1, 2, 3,$ and 4 correspond to those contained in the table labeled (B). Therefore, answer choice (B) is correct.

Graphing Functions

A function written as an equation may be plotted as a graph on a coordinate grid. Because graphs use *x* and *y* coordinates, let $y = f(x)$. See if you can plot the graph of the equation $y = 2x - 5$. The first step is to make a table of values. After that, plot each point, based on its *x* and *y* values, on a coordinate grid. Finally, connect the points on the graph.

x	y = 2x − 5	y
−2	y = 2 • −2 − 5	−9
−1	y = 2 • −1 − 5	−7
0	y = 2 • 0 − 5	−5
1	y = 2 • 1 − 5	−3
2	y = 2 • 2 − 5	−1
3	y = 2 • 3 − 5	1
4	y = 2 • 4 − 5	3
5	y = 2 • 5 − 5	5

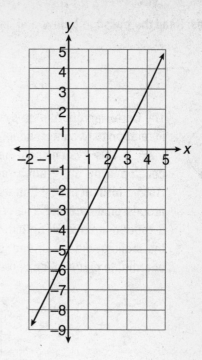

Slope

Slope describes a property of the graph of a function. **Slope** is the ratio of the rise (or vertical change) to the run (or horizontal change) as you move from one point on the graph to another. The steepness of the slope tells you how much a graph is changing vertically compared to how much the graph is changing horizontally. The formula for finding the slope of a line is $m = \dfrac{y_2 - y_1}{x_2 - x_1}$, where m is the slope and (x_2, y_2) and (x_1, y_1) are the coordinates of two different points on the line.

Figure out the slope of the line of $y = 2x − 5$. You may choose any two points, so suppose you chose (5, 5) and (3, 1) and plugged these values into the formula for slope.

$$m = \frac{y_2 - y_1}{x_2 - x_1}$$

$$m = \frac{5 - 1}{5 - 3}$$

$$m = \frac{4}{2}$$

$$m = 2$$

The slope of the line of the equation $y = 2x − 5$ is 2. If you look at the graph, you will see that this makes sense. Move along the line from left to right, and you will find that every rise (vertical change) of two boxes corresponds to a run (horizontal change) of one box.

Descartes's Fly

The Cartesian coordinate system in geometry uses numbers to locate a point in relation to two intersecting straight lines (now known as the x-axis and the y-axis). This system has many uses in the world of mathematics. It was developed by the French philosopher and mathematician René Descartes (1596–1650). It is said that as he lay sick in bed one day, he saw a fly buzzing around on the ceiling. The ceiling happened to be made of square tiles. All of a sudden, it dawned on him that he could describe the position of the fly by the ceiling tile it landed on. After this experience, he developed the coordinate plane to make it easier to describe the position of objects.

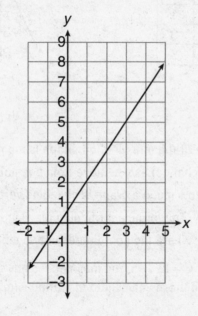

▶ The equation $2y = 3x + 1$ is shown on the graph above. What is its slope?

A $-\dfrac{3}{2}$

B $-\dfrac{2}{3}$

C $\dfrac{2}{3}$

D $\dfrac{3}{2}$

Know It All Approach

Read the question and consider the graph carefully. The formula for slope is $m = \dfrac{y_2 - y_1}{x_2 - x_1}$. To use the formula, choose two points on the graph as (x_2, y_2) and (x_1, y_1). To make your calculations easier, choose points with whole-number values. For example, for the first point, choose $(3, 5)$ and for the second point choose $(5, 8)$. Plug these values into the formula, and then solve for the slope.

$$m = \frac{y_2 - y_1}{x_2 - x_1}$$

$$m = \frac{8 - 5}{5 - 3}$$

$$m = \frac{3}{2}$$

A good way to double-check your answer is to find the slope using two different points. For example, you could use $(1, 2)$ and $(-1, -1)$.

$$m = \frac{y_2 - y_1}{x_2 - x_1}$$

$$m = \frac{-1 - 2}{-1 - 1}$$

$$m = \frac{-3}{-2}$$

$$m = \frac{3}{2}$$

The process of elimination isn't much help for this question. None of the answer choices is clearly incorrect. In fact, the incorrect answer choices are there to trick you! If you were not careful subtracting positive and negative numbers, you might have chosen (A) or (B). If you placed the *x*-values over the *y*-values, you might have chosen (B) or (C). However, if you worked carefully and double-checked your answer, you would have chosen (D) as the correct answer.

Directions: Read the passage below and answer the questions that follow.

The Seeing Eye™

Morris Frank was sixteen years old when he lost his sight. His blindness made it very difficult for him to get around on his own. He wrote a moving letter to Dorothy Eustis, a champion breeder and trainer of German shepherds, pleading with her to do something to help. Ms. Eustis invited him to visit her in Switzerland to see what could be done. Morris arrived in Switzerland in April, 1928. He stayed a number of weeks, learning to work with a German shepherd originally named Kiss. (Morris later changed her name to Buddy.)

Upon returning to America, Morris and Buddy traveled across the country to get support for the training and use of seeing-eye™ dogs. Together they showed that a blind person could walk down a street, enter a building, cross at a busy street corner, or get on a streetcar with the help of a well-trained dog. In 1929, Dorothy Eustis came to the United States. With help from Morris Frank and Buddy, she established "The Seeing Eye™," a pioneer guide-dog school in the United States.

1. The change in the number of visually impaired people with seeing-eye™ dogs in a certain country over a period of twenty years is shown in the table below.

Year	Number of Visually Impaired People with Seeing–Eye™ Dogs
1955	180,000
1960	205,000
1965	230,000
1970	255,000

If this pattern continued, in what year would the number of visually impaired people with seeing-eye™ dogs have surpassed 350,000?

A 1980
B 1985
C 1990
D 1995

2. Which of the tables shown below lists the correct values for the function $y = -2x + 5$?

x	y
−2	9
−1	7
0	5
1	3

A

x	y
−2	1
−1	3
0	5
1	7

B

x	y
−2	3
−1	5
0	7
1	9

C

x	y
−2	1
−1	−1
0	−3
1	−5

D

Directions: Read the passage below and answer the questions that follow.

Food Fight!

You will probably get in trouble if you ever throw food at other people. But you wouldn't get in trouble for starting a food fight in Buñol, a small village in Spain. At least not on the last Wednesday of August, at the peak of tomato season, when the village stages a tomato war. For two hours, thousands of people pelt each other with red, ripe tomatoes in a festival called La Tomatina. It is arguably the world's largest food fight! Every year the number of tons of tomatoes increases as does the number of participants from all over the world. Despite the potential for chaos, the festival has been a success every year without any incidents of injury.

3. Imagine a graph of the number of tomatoes thrown during La Tomatina had a slope similar to the graph below. What is the slope of the graph of an equation shown on the graph below?

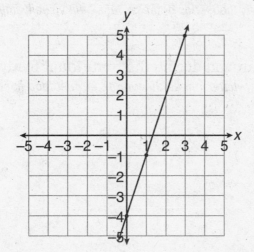

A -3

B $-\dfrac{3}{2}$

C $\dfrac{3}{2}$

D 3

Subject Review

In Chapter 8, you learned about number patterns and geometric patterns. You also learned that a function is a relationship specified by a rule that pairs up each input with a corresponding output. Furthermore, you found out that most functions may be written as equations and plotted on a graph. Lastly, you learned that the slope of a line is the ratio of the rise, or vertical change, to the run, or horizontal change, as you move from one point on the graph to another.

And, of course, you learned about Descartes, seeing-eye™ dogs, and a spectacular food fight.

What was René Descartes watching when he came up with the idea of the x- and y-coordinate graph?
A fly buzzing around on the ceiling inspired Descartes to invent the coordinate graph.

Who was the first seeing-eye™ dog?
A German shepherd named "Buddy" (originally named "Kiss") is considered the first seeing-eye™ dog. Today, thousands of these dogs help visually impaired people all over the world.

What gets tossed around during the traditional festival of La Tomatina?
Tomatoes are tossed in what might be the world's largest food fight.

Chapter 9: Solving Equations and Inequalities

Who invented the modern-day copy machine?

What is the world's largest igloo?

What is the world's longest pizza delivery?

Linear Equations

In the last chapter, you learned that most functions could be written as equations. Moreover, an equation can be plotted on a coordinate graph. All the graphs you have worked with thus far have been straight lines. A **linear equation** is any equation whose graph is a straight line. Examples of linear equations are $x + 4 = 11$, $2x - 3 = 8$, and $4x + 5 = x - 4$.

Math teachers did not come up with the idea of equations just to make life more complicated. Equations can help you solve real problems. Whether or not you realize it, you use equations every day. Look at the sample question below.

▶ Moesha has her own telephone. She pays $24.00 per month plus $0.15 for each local call. Long-distance calls are extra. Last month, her bill was $41.65. It included $8.35 in long-distance calls. How many local calls did Moesha make last month?

This is a word problem. To solve a word problem, you need to write an algebraic expression that states the same situation as described in words. What do you know? You know the total bill ($41.65), the monthly charge ($24.00), the long-distance charges ($8.35), and the charge for each local call ($0.15). What are you trying to find out? Let n represent the number of local calls Moesha made last month. The charges for local calls may be expressed as $0.15n$. You know that this charge is separate from the flat monthly charge and the long-distance charges. Therefore, you can write an equation that summarizes the situation.

Total bill = Monthly charge + Local charges + Long-distance charges

$41.65 = $24.00 + $0.15n + $8.35

Now what? You need to solve for the unknown variable, n. The key to doing this is to isolate the variable on one side of the equation. Remember that, as long as you do the same thing to both sides, an equation remains true.

$41.65 = 24.00 + 0.15n + 8.35$ Combine like terms.

$41.65 = 32.35 + 0.15n$ Subtract 32.35 from both sides.

$9.30 = 0.15n$ Divide both sides by 0.15.

$62 = n$

That's the answer! Moesha made sixty-two local calls last month. Aren't equations great? Of course, some equations are more complicated than others. Try the next sample problem.

▶ Solve $5x - 3 = 2x + 9$. Show all the steps you took to determine the solution.

Remember that the key to solving an equation is to isolate the variable on one side of the equation.

$5x - 3 = 2x + 9$ Add $(-2x)$ to both sides.

$3x - 3 = 9$ Add $+ 3$ to both sides.

$3x = 12$ Divide both sides by 3.

$x = 4$

Systems of Equations

Some real-life situations require more than one equation. A **system of equations** is a set of two or more equations. Why would you need two equations? To find out, study the sample question below.

▶ Ben has 16 coins in his pocket that add up to $2.00. Every coin is either a quarter or a nickel. How many quarters does Ben have?

How can you find the answer? One way is by trial and error.

1 quarter and 15 nickels = $1 \cdot \$0.25 + 15 \cdot \$0.05 = \$1.00$ No.

2 quarters and 14 nickels = $2 \cdot \$0.25 + 14 \cdot \$0.05 = \$1.20$ No.

3 quarters and 13 nickels = $3 \cdot \$0.25 + 13 \cdot \$0.05 = \$1.40$ No.

Etc., etc., etc.

As you can see, this is rather time-consuming. Eventually, you'll arrive at the correct answer, but is there a better way? Yes there is! Use a system of equations!

Let q represent the number of quarters, and let n represent the number of nickels. Now write two equations that represent the situation described in the question.

$$q + n = 16$$

$$\$0.25q + \$0.05n = \$2.00$$

Okay, that's terrific. Now what? Write an expression that gives the value of n in terms of q. If $q + n = 16$, then $n = 16 - q$ (you add "$-q$" to both sides). Now you can substitute "$16 - q$" for n in the second equation.

$0.25q + 0.05n = 2.00$	Substitute "$16 - q$" for n; clear parentheses.
$0.25q + 0.05(16 - q) = 2.00$	Multiply the parentheses.
$0.25q + 0.80 - 0.05q = 2.00$	Combine like terms.
$0.20q + 0.80 = 2.00$	Add (-0.80) to both sides.
$0.20q = 1.20$	Divide both sides by 0.20
$q = 6$	

The correct answer is that Ben has six quarters. The question didn't ask it, but do you know how many nickels he has? That's easy to figure out. Just go back to the first equation.

$q + n = 16$	Substitute 6 for q.
$6 + n = 16$	Add -6 to both sides.
$n = 10$	

So, $q = 6$ and $n = 10$. Therefore, Ben has 6 quarters and 10 nickels. How can you double-check your answers? Just plug them into the second equation.

$$0.25q + 0.05n = 2.00$$

$$0.25 \cdot 6 + 0.05 \cdot 10 = 2.00$$

$$1.50 + 0.50 = 2.00$$

$$2.00 = 2.00$$

This method of solving a system of equations is called the **substitution** method. Sometimes, you can use a different method if the situation allows. Check out the next sample question.

▶ Maria and Alex ate lunch at the pizzeria. Maria ordered one slice of pizza and one soda. Alex ordered three slices of pizza and two sodas. Maria's bill was $2.30, and Alex's bill was $6.10. What was the price of one slice of pizza? What was the price of one soda?

Let p represent the price of one slice of pizza, and let s represent the price of one soda. Next, write an equation for each person's bill.

$p + s = \$2.30 =$ Maria's bill

$3p + 2s = \$6.10 =$ Alex's bill

Wouldn't it be convenient if one of the unknown variables (p or s) disappeared? Then you could solve for one variable as you did at the beginning of the chapter. Well, there is a way to make this happen. Remember, an equation remains true as long as you do the same thing to both sides. Multiply both sides of Maria's equation by 2.

$2(p + s) = 2(2.30) \qquad 2p + 2s = 4.60$

Then subtract Alex's equation from Maria's new equation.

$2p + 2s = 4.60$

$-3p + 2s = 6.10$

$-1p = -1.50$

$p = 1.50$

Notice what happens. The terms with "$2s$" cancel each other. You are left with $-1p = -\$1.50$, which easily translates (multiply both sides by -1) into $p = \$1.50$.

Therefore, the price of one slice of pizza was $1.50. What was the price of one soda? You can figure out this answer by referring back to either of the two original equations.

$p + s = 2.30$

$1.50 + s = 2.30$

$s = 0.80$

Therefore, the price of one soda is $0.80. Double check your answer by substituting both answers into the other equation.

$3p + 2s = 6.10$

$3(1.50) + 2(0.80) = 6.10$

$4.50 + 1.60 = 6.10$

$6.10 = 6.10$

The answer checks, so it must be correct. This method of solving a system of equations is called the **addition** method.

Inequalities

An **inequality** is a mathematical sentence that contains one of the following symbols: $>, <, \geq, \leq,$ or $\neq$. Look at some examples of inequalities, and notice how to express them in words. Then try the sample questions below.

$x > 5$	x "is greater than" 5
$n < -1$	n "is less than" -1
$y \geq 3x$	y "is greater than or equal to" $3x$
$x + 2y \leq 8$	$x + 2y$ "is less than or equal to" 8
$b \neq 0$	b "is not equal to" 0

▶ Which of the inequalities listed below represents the sentence, "5 added to a number, n, is greater than or equal to 8"?

A $5 \geq n + 8$
B $n + 5 \leq 8$
C $n + 5 > 8$
D $n + 5 \geq 8$

(D) is the correct answer choice. Solving inequalities is similar to solving equations. The key is to isolate the variable on one side of the inequality. Check out another sample question.

▶ Solve: $2x - 6 < 2$

Follow the same steps you would to solve for x in an equation.

$2x - 6 < 2$	Add $+6$ to both sides.
$2x < 8$	Divide both sides by the coefficient of x.
$x < 4$	

The correct answer is $x < 4$, or "x is less than 4." Therefore, any number less than 4 makes the inequality true.

There is, however, one **very important** difference when working with inequalities as opposed to equations. To understand that difference, look at another sample question.

▶ Solve: $-6x < 12$.

Naturally, as a first step, you should divide both sides by -6, the coefficient of x.

$$-6x < 12$$
$$x < -2$$

Piece of cake, right? Unfortunately, this answer is **incorrect**! How do you know? Try substituting a number less than -2 in place of x. Try -3.

$$-6x < 12$$
$$-6 \cdot -3 < 12$$
$$18 < 12$$

This is false. Is eighteen less than twelve? Of course not! Obviously, something is wrong here. Now you will learn the one **very important** difference referred to earlier. Whenever you multiply or divide an inequality by a **negative** number, it changes the **direction** of the inequality. Now try to solve the sample question.

$$-6x < 12$$
$$x > -2$$

You had to change the **direction** of the symbol! Now you have the correct answer: $x > -2$, or "x is **greater** than -2." Double-check your answer by substituting a number greater than -2 in place of x. Try -1.

$$-6x < 12$$
$$-6 \cdot -1 < 12$$
$$6 < 12$$

True!

Graphing Inequalities

If you can graph an equation, you can graph an inequality. See if you can figure out how to graph the following inequality.

$$y \le 3x - 1$$

The first thing you need to do is graph the sentence as if it contained an "equals" sign ($=$). As you may recall from the last chapter, the first step is to make a table of values. After that, plot each point, based on its x and y values, on a coordinate grid. Finally, connect the points on the graph. They should form a straight line.

x	y = 3x − 1	y
−2	y = 3 • −2 − 1	−7
−1	y = 3 • −1 − 1	−4
0	y = 3 • 0 − 1	−1
1	y = 3 • 1 − 1	2
2	y = 3 • 2 − 1	5
3	y = 3 • 3 − 1	8
4	y = 3 • 4 − 1	11
5	y = 3 • 5 − 1	·14

The line should be drawn **solid** if the inequality is ≤ or ≥. Draw the line **dashed** if the inequality is < or >. For the next step, you should pick a point not on the line to use as a test point. A good test point is (0, 0), provided it is not on the line.

$y \leq 3x - 1$

$0 \leq 3 \cdot 0 - 1$

$0 \leq -1$

False!

If the test point makes the inequality true, shade that side of the line. If the test point makes the inequality false, shade the opposite side of the line. Thus, the graph of the inequality $y \leq 3x - 1$ should look like the figure shown at right. All points located in the shaded portion of the graph make the inequality true.

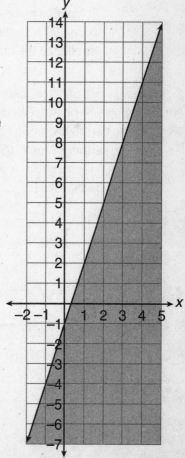

Directions: Read the passage below and answer the question that follows.

Mr. Copy Cat

Physicist Chester F. Carlson is known as the father of xerographic printing. He was driven by a need for some way to make multiple copies of scale drawings of inventions so they could be submitted for patents. Working out of his apartment, he experimented with different methods for automatically reproducing text and diagrams. While others thought that chemical or photographic solutions were the answer, Carlson concentrated his efforts on electrostatics.

In 1938, he succeeded in obtaining his first "dry-copy" and the first of many patents. He demonstrated his invention to more than twenty companies before the Battelle Development Corporation became interested. In 1947, the Haloid Company—renamed Xerox Corporation—purchased the commercial rights to his electrostatic copying process. Eleven years later, Xerox introduced the first office copier.

▶ A company sells toner cartridges for copy machines. This month, it has two specials for those who buy toner cartridges in bulk. The first special includes 12 brand-name cartridges and 20 generic cartridges for $1,728. The second special includes 8 brand-name cartridges and 12 generic cartridges for $1,092. This situation is represented by the system of equations below, in which b represents the price of one brand-name cartridge and g represents the price of one generic cartridge.

$$12b + 20g = \$1{,}728$$
$$8b + 12g = \$1{,}092$$

What is the price of one brand-name toner cartridge?

_____

Know It All Approach

You need to solve the system of equations to determine the price of one brand-name toner cartridge.

First, read the question carefully and make note of the important information. You don't need to write down the numbers given in the question because they are already contained in the system of equations. You do need to know what b and g represent. In addition, you need to know that you're trying to find the value of b, that is, the price of one brand-name toner cartridge.

Next, calculate the answer. Earlier, you learned about two methods of solving a system of equations: substitution and addition. Examine the two equations carefully. Try using the substitution method. Write an expression that states the value of g in terms of b.

$$8b + 12g = \$1,092$$

$$8b = \$1,092 - 12g$$

$$b = 136.5 - 1.5g$$

This method would lead to the correct answer, but it is difficult to work with because of the decimals. Next, try using the addition method. Notice that you cannot cancel any terms in the equations as they now stand. Look at the terms $8b$ and $12b$. What number do both 8 and 12 go into evenly? If you said 24, you're correct. Thus, you should multiply both sides of the first equation by 2, and both sides of the second equation by 3. Then subtract to get rid of the terms that contain b.

$$2 \bullet (12b + 20g = 1,728) = 24b + 40g = 3,456$$

$$3 \bullet (8b + 12g = \$1,092) = 24b + 36g = 3,276$$

$$(24b + 40g = 3,456) - (24b + 36g = 3,276)$$

$$4g = 180$$

$$g = 45$$

Now you know that g, the price of a generic cartridge, equals \$45. However, you need to find b. Go back to either of the original equations, and substitute \$45 for g.

$$8b + 12g = 1,092$$

$$8b + 12 \bullet 45 = 1,092$$

$$8b + 540 = 1,092$$

$$8b = 552$$

$$b = 69$$

Therefore, you know that *b*, the price of a brand-name cartridge equals $69.

After you have completed your calculations, check your answer. The best way to do this is to substitute your answers into the other equation.

$$12b + 20g = 1,728$$

$$12 \cdot 69 + 20 \cdot 45 = 1,728$$

$$828 + 900 = 1,728$$

$$1,728 = 1,728$$

Finally, write the answer to the question clearly in the space provided.

$b = \$69$

Directions: Read the passage below and answer the questions that follow.

A Cool Vacation Spot

Imagine staying in a hotel and having the temperature in your room drop below 23° F (−5° C). Most likely, you would pick up the phone and complain to the manager. However, guests at the Ice Hotel in Sweden wouldn't have it any other way. It is the world's largest igloo!

Every October, builders start rebuilding the hotel by blasting thirty thousand tons of snow onto a molded frame. Next, it is reinforced with ten thousand tons of ice sheets and ice blocks. Lastly, the ice sculptors come in to carve out doors, windows, tables, chairs, and beds. By the middle of December, the hotel is opened to the public.

Staying at the Ice Hotel is the "in" thing to do in Europe. Many of the regular guests are famous politicians, entertainers, supermodels, and sports figures. Even royalty has been known to check in. The hotel supplies its guests with padded suits, gloves, thermal boots, and warm reindeer skins. In May, the Ice Hotel begins to melt and run into a nearby river. People can't wait until construction begins anew the following October!

Know It All! High School Math

1. You can determine the amount of heat needed to melt a block of ice using the formula: $Q = mCT$, where Q is the amount of heat, m is the mass, C is the specific heat, and T is the temperature change. Which of the following expressions shows the value of m in terms of the other three variables?

   A $m = C - T$

   B $m = \dfrac{Q}{CT}$

   C $m = \dfrac{CT}{Q}$

   D $m = Q - C - T$

2. Solve: $5(3x - 8) = 5x - 10$

   A $x = -6$

   B $x = -5$

   C $x = -\dfrac{1}{5}$

   D $x = 3$

3. What is the solution set for the system of equations shown below?

   $y = 3x - 6$

   $2x + y = 19$

   A   (2, 0)

   B   (4, 6)

   C   (5, 9)

   D   (9, 1)

4. Solve for x: $2(2x - 1) - 5x < -3$

   A $x < -3$

   B $x < -1$

   C $x > 1$

   D $x > 5$

Special Delivery!

On March 22, 2001, Butlers Pizza, located in Cape Town, South Africa, received an order for one pizza pie from Corme Krige, captain of the Fedsure Stormers rugby team. He was calling from Sydney, Australia! Bernard Jordaan traveled a distance of 11,042 kilometers (6,861 miles) to deliver the pizza in person. It is not known whether the pizza was still warm when it arrived.

Eddie Fishbaum, owner of the Jerusalem 2 pizza parlor in New York City, once hand-delivered a pizza to Osaka, Japan. That's a distance of 11,126 kilometers (6,914 miles), and the pizza cost $7,000!

5. The circumference, C, of a pizza pie may be found using the formula $C = 2\pi r$, in which r is the radius and π equals 3.14. The area, A, of a pizza pie may be found using the formula $A = \pi r^2$. What is the circumference of a pizza pie with an area of 314 square inches?

    A   10.0 in.
    B   20.0 in.
    C   31.4 in.
    D   62.8 in.

6. Rowena sold tickets for two performances of the school play. For Friday night's performance, she sold twenty-five floor-level seats and ten balcony seats, and collected $350. For Saturday night's performance, she sold twenty-two floor-level seats and twelve balcony seats, and collected $324. This scenario is represented by the following system of equations:

$25f + 10b = \$350$

$22f + 12b = \$324$

In these equations, f represents the cost of one floor-level seat and b represents the cost of one balcony seat. Solve the system of equations to find the cost of one floor-level seat.

_____

7. Solve for n: $3n + 10 = 4$

A   −6
B   −2
C   2
D   6

Subject Review

In Chapter 9, you learned how to solve linear equations. You also learned how to solve systems of equations. In addition, you learned how to solve inequalities. You were told about the one important exception when solving inequalities. Lastly, you learned how to graph inequalities.

You are also the proud owner of a few new facts.

Who invented the modern-day copy machine?
Chester F. Carlson is given credit for the useful invention.

What is the world's largest igloo?
If it's winter, then it's Sweden's Ice Hotel. But the popular tourist attraction melts every spring!

What is the world's longest pizza delivery?
Eiji Bando, a Japanese celebrity, ordered a pizza from a New York pizza shop. Eddie Fishbaum delivered a pie 11,126 kilometers (6,914 miles) from New York City to Osaka, Japan.

Chapter 10: Graphing Equations

Who made important contributions to the discovery of the structure of DNA but was never given the proper recognition in her lifetime?

Who invented carbonated water?

What roller coaster broke four world records?

Graphing Linear Equations

A **linear equation** is any equation whose graph is a straight line. In Chapter 8, you learned how to graph a function that could be expressed as a linear equation. In Chapter 9, you learned about systems of equations. So far, you know of two methods to solve a system of equations—the substitution method and the addition method. There is another method—the graphing method. Consider the sample question below.

▶ What is the solution set for the system of equations shown below?

$$y = 2x + 5$$

$$x + y = 2$$

A $(-3, 1)$
B $(-1, 3)$
C $(1, -3)$
D $(3, 1)$

You already know how to solve this system of equations using either the substitution method or the addition method. Now, try solving it using the graphing method. The first step is to graph one of the equations. Start with $y = 2x + 5$. As you know, you will need to make a table of values, plot each point on a coordinate grid, and connect the points.

x	y = 2x + 5	y
−3	y = 2 • −3 + 5	−1
−2	y = 2 • −2 + 5	1
−1	y = 2 • −1 + 5	3
0	y = 2 • 0 + 5	5
1	y = 2 • 1 + 5	7
2	y = 2 • 2 + 5	9
3	y = 2 • 3 + 5	11

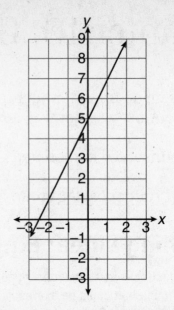

The next step is to graph the second equation on the same coordinate grid. Again, you will need to make a table of values, plot each point, and then connect all the points. It would be difficult to make a table of values for the equation $x + y = 2$ in its present form. Rewrite the equation to isolate y on one side. Then, make your table of values.

$x + y = 2$ Add $-x$ to both sides.

$y = 2 - x$

x	y = 2 − x	y
−3	2 − (−3)	5
−2	2 − (−2)	4
−1	2 − (−1)	3
0	2 − 0	2
1	2 − 1	1
2	2 − 2	0
3	2 − 3	−1

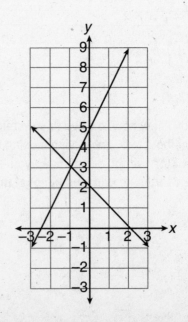

Know It All! High School Math

Examine both graphs. Do you notice anything? They intersect! The point where they intersect is the solution set for the system of equations. That point, $(-1, 3)$, has the values for x and y that make both equations true. Therefore, answer choice (B) is correct.

How could you double-check your answer? The easiest way would be to solve the system of equations using another method. Then see if you get the same answer. Try the substitution method.

$$y = 2x + 5$$

$$x + y = 2$$

The first equation tells you that $y = 2x + 5$. Substitute this expression for y in the second equation.

$$x + y = 2$$

$$x + 2x + 5 = 2$$

$$3x + 5 = 2$$

$$3x = -3$$

$$x = -1$$

Lastly, go back to either of the original equations. Substitute -1 in place of x.

$$x + y = 2$$

$$-1 + y = 2$$

$$y = 3$$

The solution set is $x = -1$, $y = 3$, or $(-1, 3)$. This is the same answer you got using the graphing method. Your answer checks, so you know it is correct.

Graphing Quadratic Equations

Did you ever notice that linear equations never contain exponents? **Quadratic equations** do. They always have a variable that is raised to the power of 2. Many natural phenomena follow mathematical rules that contain variables raised to the power of 2. For example, the area of a circle is found using the formula $A = \pi r^2$, where A is the area, π equals 3.14, and r is the radius of the circle. $A = \pi r^2$ is a quadratic equation. What does the graph of a quadratic equation look like?

Make a table of values for this equation and then use the values to make a graph. Label the y-axis "Area (square inches)" and the x-axis "Radius (inches)."

r	$A = \pi r^2$	A
1	$A = 3.14 \cdot 1^2$	3.14
2	$A = 3.14 \cdot 2^2$	12.56
3	$A = 3.14 \cdot 3^2$	28.26
4	$A = 3.14 \cdot 4^2$	50.24
5	$A = 3.14 \cdot 5^2$	78.50

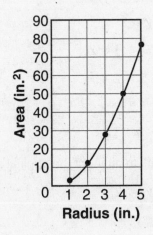

How about that—the graph is a **curved** line! That's what a quadratic equation looks like.

Now try the following sample question with a quadratic equation.

▶ Make a graph of the quadratic equation $y = x^2 - 3x - 10$. Make a table of values from $x = -4$ to $x = 7$. Plug the values into the equation to find the values for y. Then plot the points on a coordinate grid and connect them with a smooth curved line.

x	$y = x^2 - 3x - 10$	y
−4	$y = (-4)^2 - 3 \cdot -4 - 10$	18
−3	$y = (-3)^2 - 3 \cdot -3 - 10$	8
−2	$y = (-2)^2 - 3 \cdot -2 - 10$	0
−1	$y = (-1)^2 - 3 \cdot -1 - 10$	−6
0	$y = (-0)^2 - 3 \cdot 0 - 10$	−10
1	$y = (-1)^2 - 3 \cdot 1 - 10$	−12
2	$y = (-2)^2 - 3 \cdot 2 - 10$	−12
3	$y = (-3)^2 - 3 \cdot 3 - 10$	−10
4	$y = (-4)^2 - 3 \cdot 4 - 10$	−6
5	$y = (-5)^2 - 3 \cdot 5 - 10$	0
6	$y = (-6)^2 - 3 \cdot 6 - 10$	8
7	$y = (-7)^2 - 3 \cdot 7 - 10$	18

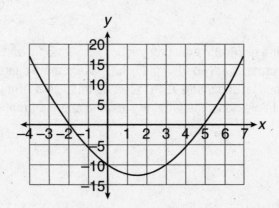

Okay, so it's a curve. Do you remember that in Chapter 7 you learned how to factor a polynomial? Try factoring the trinomial on the right side of this quadratic equation.

▶ Factor: $x^2 - 3x - 10 = 0$

To get the first term of x^2, the only possible factors are x and x. So you start with $(x\ \ \ )(x\ \ \ )$.
To get the last term of -10, there are many possible factors. You need to find the combination that produces the middle term of $-3x$.

$+ 10$ and -1	$(x + 10)(x - 1)$	$10x - 1x = 9x$	Wrong.
$- 10$ and $+ 1$	$(x - 10)(x + 1)$	$-10x + 1x = -9x$	Wrong.
$+ 5$ and -2	$(x + 5)(x - 2)$	$5x - 2x = 3x$	Wrong.
$- 5$ and $+ 2$	$(x - 5)(x + 2)$	$-5x + 2x = -3x$	Right!

Therefore, the factors are $(x - 5)(x + 2)$. If you plug in these two factors in place of the trinomial, you get: $(x - 5)(x + 2) = 0$. Now, see the beauty of algebra? If the product of two factors equals zero, then one or both of the factors must equal zero. So all you have to do is make each of the factors equal to zero. Then, solve for x.

$$x - 5 = 0 \qquad x + 2 = 0$$
$$\underline{+ 5\ +5} \qquad \underline{- 2\ -2}$$

$$x = 5 \qquad\quad x = -2$$

Now look at the graph of the quadratic equation $y = x^2 - 3x - 10$. At which two points does the graph cross the y-axis? It crosses at $(-2, 0)$ and $(5, 0)$. These are the same values you got for x when you factored the trinomial and made each factor equal to zero. You just solved a quadratic equation using two different methods: the factoring method and the graphing method!

Directions: Read the passage below and answer the question that follows.

Unsung Hero of Science

The 1962 Nobel Prize for Physiology and Medicine was awarded to James Watson, Francis Crick, and Maurice Wilkins for their landmark research on DNA in which they discovered that DNA molecules have a "double helix" structure. Notably absent from the ceremony, due to her premature death at age thirty-seven, was Rosalind Franklin. Her contribution to the discovery of the structure of DNA was crucial. She was an expert in X-ray diffraction, a procedure that uses X-rays to create specialized images. She led the way in the use of this procedure for analyzing large biological molecules. Early in her research, she realized an important aspect of the structure of DNA molecules.

Without Rosalind Franklin's permission, her colleague, Maurice Wilkins, shared her data with Watson and Crick. When they saw her excellent X-ray diffraction photographs, they were convinced that the DNA molecule had to have a double helix structure. In March 1953, they reported their findings in a published scientific paper. Eventually, after Franklin died of cancer in 1958, Watson and Crick agreed that she had made important contributions to the discovery of the structure of DNA.

Rosalind Franklin

▶ Scientists were studying the DNA of beetles. Five years ago, they had one thousand beetles in their laboratory. The beetle population grew according to the equation $y = x^2 + 2x + 1$. In this equation, y is the beetle population in thousands (1,000s) and x is the number of years that have passed. Make a table of values from $x = 0$ to $x = 5$, and then graph the equation.

x	$y = x^2 + 2x + 1$	y

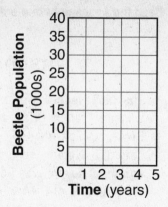

Know It All Approach

To begin, read the question carefully, and make sure you understand the information that has been provided. Also, make sure you understand that the question is asking you to do two things: make a table of values and make a graph of those values.

Next, make a table with three columns and seven rows as shown below. Plug each x-value into the equation to figure out each y-value. Once you have figured out all the values for x and y, plot the points on a coordinate grid and connect them with a smooth line. Label the x-axis "Time (years)," and the y-axis "Beetle Population (1,000s)."

x	$y = x^2 + 2x + 1$	y
0	$y = 0^2 + 2 \cdot 0 + 1$	1
1	$y = 1^2 + 2 \cdot 1 + 1$	4
2	$y = 2^2 + 2 \cdot 2 + 1$	9
3	$y = 3^2 + 2 \cdot 3 + 1$	16
4	$y = 4^2 + 2 \cdot 4 + 1$	25
5	$y = 5^2 + 2 \cdot 5 + 1$	36

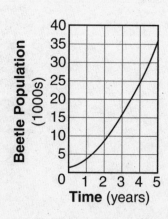

Remember to double-check your answer. A good way to recheck a graph of a quadratic equation is to examine its shape. If it is not a smooth curve, you may have made an error in calculating one or more of the y-values. Check that you have completely answered the question and that your answer is written clearly.

Directions: Read the passage below and answer the questions that follow.

The Soda Story

In 1772, Joseph Priestley, the scientist who discovered oxygen, figured out a way to trap carbon dioxide in water. Carbon dioxide is an odorless, colorless gas that we breathe out when we exhale. When dissolved in water, carbon dioxide produces a "sparkling" effect as it escapes when the water is poured into a glass. Soda was born! To manufacture a carbonated soft drink, bottling companies start with ordinary tap water. They treat the water to remove impurities and some of the naturally occurring minerals. Then they bubble carbon dioxide gas into a pressurized container filled with the treated water. Finally, they add sweeteners and flavoring to produce your favorite soda. Many people agree that carbonation makes soft drinks taste better and adds to the feeling of having your thirst quenched.

1. A restaurant conducting a survey about soda found that there were relationships between the temperature outside (x) and how many glasses of soda were sold (y) for an orange soda and a cola soda. The system of equations below shows the information found by the surveys. Solve the system of equations shown below by **graphing**. **Show all work**, including tables, graphs, and the solution set.

    Orange Soda: $y = 3x$

    Cola Soda: $x + y = 8$

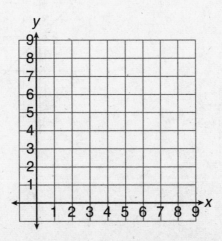

2. Which of the following graphs represents the equation $y = 2x + 3$?

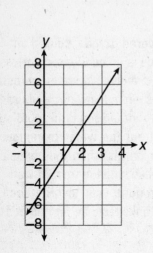

A

B

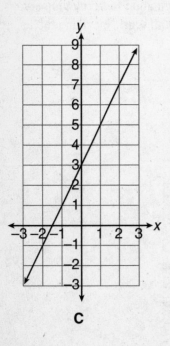

C

D

The Steel Dragon

Are you a thrill seeker? Then this ride is for you! The Steel Dragon officially opened on August 1, 2000, at Nagashima Spaland in Mie, Japan. At the opening ceremonies, Christopher Irwin, CEO of Guinness World Records, presented park president Mr. Otani with four certificates for world records for a roller coaster. They included the Longest Coaster in the World (2,479 meters, or 8,133.60 feet), Fastest Coaster in the World (153.2 kilometers per hour, or 95 miles per hour), Tallest First Drop (93.5 meters, or 306.77 feet), and Height of First Three Hills (318 feet, 252 feet, 210 feet).

After being towed up to a height of 95 meters (311.7 feet), the ride begins with a speedy fall followed by two more hills and dips. In the first minute, you travel over 1,340 meters (4,400 feet), average more than 90 kilometers per hour (55 miles per hour), and reach speeds more than 105 kilometers per hour (65 miles per hour) five times. At the top of the third hill, a series of spiral twists awaits you. The return loop has eight more bumps, seven more dips, and two tunnels! When you finish the 2,479 meter (8,133 feet) round trip, you have dropped almost a total of 365 meters (1,200 feet).

The Steel Dragon's first drop

3. The distance fallen by a free-falling body can be determined using the equation $y = 16x^2$, where y represents the distance fallen in feet and x represents the time in seconds of free fall. Make a table of values for $x = 0$ to $x = 5$, and then graph the equation. Lastly, answer this question: How far would a skydiver free fall in five seconds?

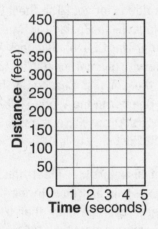

Distance (feet) vs. Time (seconds)

_____

Subject Review

In Chapter 10, you learned how to solve a system of equations by graphing. You also learned that a quadratic equation is one in which a variable is raised to the second power. Lastly, you learned how to solve a quadratic equation by graphing.

Who made important contributions to the discovery of the structure of DNA but was never given the proper recognition in her lifetime?
Rosalind Franklin made important contributions to the discovery of the structure of DNA. She eventually received some recognition after her death from cancer.

Who invented carbonated water?
Joseph Priestley figured out how to trap carbon dioxide in water. Soda manufacturers trap carbon dioxide in the water to create the fun bubbles we associate with a satisfying soda.

What roller coaster broke four world records?
The Steel Dragon in Japan has been recognized by the Guinness World Records as being the longest and fastest roller coaster, and has received four certificates for world records. The roller coaster is 2,479 meters long and reaches speeds of 153.2 kilometers per hour!

Chapter 11: Polygons and Solids

What is "Punkin' Chunkin'"?

Who invented the zipper?

What was the shortest war in world history?

Polygons

A **polygon** is a closed figure formed by three or more line segments. Some common polygons are shown below.

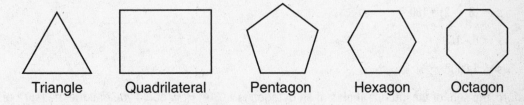

| Triangle | Quadrilateral | Pentagon | Hexagon | Octagon |

Polygons are named according to the number of sides they have. "Tri" means three, "quad" means four, "penta" means five, and so on. If all sides of a polygon are congruent, that is, equal in length, it is called a **regular** polygon.

By drawing as many diagonals as you can, you can divide any polygon into a number of triangles. The diagram below shows a hexagon and an octagon divided into triangles.

You will discover that the number of triangles you can form from one point inside a polygon equals **two less** than the number of sides. Why is this useful? You probably know that the sum of the angles inside a triangle equals 180°. By dividing a polygon into triangles, you can calculate the sum of its interior angles. Because the hexagon can be divided into four triangles, the sum of its interior angles equals 720° (4 • 180).

What is the sum of the interior angles of an octagon? The general formula is $s = (n - 2) \bullet 180°$, in which s is the sum of the interior angles and n is the number of sides.

$$s = (n - 2) \bullet 180$$

$$s = (8 - 2) \bullet 180$$

$$s = 6 \bullet 180$$

$$s = 1{,}080°$$

Therefore, the sum of the interior angles of an octagon is 1,080°. How could you find the measure of **each** interior angle? This can only be done for regular polygons. Check out the sample question on the following page.

▶ What is the measure in degrees of each interior angle of the regular pentagon shown below?

The first step is to find the sum of all the interior angles of the pentagon.

$s = (n - 2) \cdot 180$

$s = (5 - 2) \cdot 180$

$s = 3 \cdot 180$

$s = 540°$

The sum of the interior angles equals 540°. Since it is a regular pentagon, not only are all the sides congruent, but so are the angles. Therefore, each interior angle measures 108° (540 ÷ 5).

Quadrilaterals

Many of the shapes you see every day are quadrilaterals. A **quadrilateral** is a four-sided polygon. According to the formula discussed earlier, the sum of its interior angles equals 360°.

Quadrilaterals are classified according to the relationships among their sides and angles. Examine the diagram on the next page of the different types of quadrilaterals.

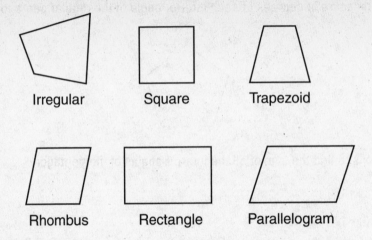

Irregular   Square   Trapezoid

Rhombus   Rectangle   Parallelogram

Notice that an **irregular** quadrilateral has no parallel sides or congruent angles.

A **rectangle** has four right angles (that is, each measures 90°), with opposite sides parallel and congruent.

A **square** is a special rectangle with all four sides congruent.

A **parallelogram** is a slanted rectangle with opposite angles congruent and opposite sides parallel and congruent.

A **rhombus** is a special parallelogram with all four sides congruent. The oddball in the group, besides the irregular quadrilateral, is the trapezoid.

A **trapezoid** has two sides parallel (called the bases) and two sides that are not parallel.

There are thousands of ways a test can ask you questions about quadrilaterals. Try a few sample questions on the following pages.

▶ In quadrilateral *ABCD*, $\overline{AD} \cong \overline{CD}$ and $\overline{AB} \cong \overline{BC}$. Is quadrilateral *ABCD* a parallelogram?

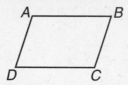

A   No, all angles must be congruent.
B   Yes, opposite sides are congruent.
C   No, both pairs of opposite sides must be congruent.
D   Yes, two pairs of sides are congruent.

The question does not mention angles, so eliminate answer choice (A). Sides *AD* and *CD* are congruent, but they are not opposite each other. The same is true of sides *AB* and *BC*, so you can throw out answer choice (B), too. Having two pairs of sides congruent does not necessarily mean that the opposite sides are congruent, so get rid of answer choice (D). Answer choice (C) is correct.

▶ Figure *EFGH* is an isosceles trapezoid. What is the measurement of ∠*G*? (Hint: In an isosceles trapezoid, the two nonparallel sides and the base angles are congruent.)

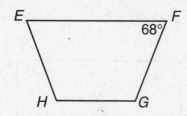

A   68°
B   112°
C   136°
D   224°

Because *EFGH* is an isosceles trapezoid, $\angle F \cong \angle E$, and $\angle G \cong \angle H$. The sum of all the interior angles equals 360°. Let *x* represent the measure of $\angle G$, then add all the angles.

$$68 + 68 + x + x = 360$$

$$136 + 2x = 360$$

$$2x = 224$$

$$x = 112$$

The measure of $\angle G$ equals 112°, so answer choice (B) is correct.

▶ A square is a

A parallelogram
B rectangle
C rhombus
D all of the above

The opposite sides of a square are parallel, so it is a parallelogram. A square has four right angles, so it is a rectangle. All four sides of a square are congruent, so it is a rhombus. Answer choice (D) is correct.

Use the rectangle below to answer the sample question on the next page.

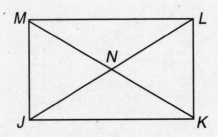

▶ $\overline{MK}$ and $\overline{JL}$ are diagonals of rectangle *JKLM* that intersect at *N*. If $\overline{MK}$ = $4n + 9$ and $\overline{JL}$ = $6n - 5$, what is the length of $\overline{MK}$? (Hint: The diagonals of a rectangle are congruent.)

   A 7
   B 13
   C 28
   D 37

Because the diagonals of a rectangle are congruent, $\overline{MK} \cong \overline{JL}$.

$$4n + 9 = 6n - 5$$

$$9 = 2n - 5$$

$$14 = 2n$$

$$7 = n$$

Don't make the mistake of thinking you're finished! You have found the value of *n*, but you still need to find the length of $\overline{MK}$. It is given in the question that $\overline{MK}$ = $4n + 9$.

$$4n + 9 = 4 \cdot 7 + 9 = 28 + 9 = 37.$$

The length of $\overline{MK}$ is 37. Therefore, answer choice (D) is correct.

Solids

A **solid** is a three-dimensional figure. Most of the objects you see every day are solids.

Check out the diagram below that shows some of the different types of solids.

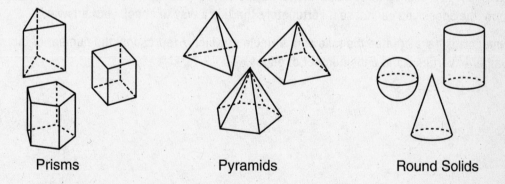

     Prisms   Pyramids   Round Solids

Round solids do not have faces or edges. You will learn more about round solids in a later chapter. The other types of solids, prisms and pyramids, are polyhedra (plural of polyhedron). A **polyhedron** is a three-dimensional figure with all flat surfaces. A prism has two parallel and congruent bases. All the other faces on a prism are rectangles. A pyramid has a polygon for a base. All the other faces on a pyramid are triangles that meet at the top.

Mathematicians have discovered some interesting relationships among these solids. Specifically, there is a mathematical relationship among the number of faces, edges, and vertices (plural of vertex). Examine the diagram below, and then try the sample question.

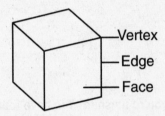

▶ How many edges does the cube shown above have? (Hint: an edge is the line along which two faces intersect.)

A 4
B 6
C 12
D 16

One way to answer this question is to count the edges. However, it can be misleading when a flat picture tries to represent a three-dimensional object. In your mind, you would have to make sure you count the edges you cannot see. Fortunately, there is a way to check your answer.

Mathematicians have created the following formula in which F represents the number of faces, V the number of vertices, and E the number of edges.

$$F + V - E = 2$$

How many faces does a cube have? Of course, you know it has six faces, just like the dice you roll when you play Monopoly or some other game. How many vertices does a cube have? A vertex is a point where three faces intersect. You can think of the vertices of a cube as its corners. There are four vertices on the top, and four more on the bottom. Thus, a cube has eight vertices. Now, answer the question using the formula.

$$F + V - E = 2$$

$$6 + 8 - E = 2$$

$$E = 12$$

A cube has twelve edges. Answer choice (C) is correct.

Directions: Read the passage below and answer the question that follows.

Punkin' Chunkin'

 What is Punkin' Chunkin'? It is a contest in which teams attempt to launch a pumpkin as far as they can. The idea started in Georgetown, Delaware, in 1986, when three friends, just joking around, challenged each other. The object was to see who could design a machine that would throw a pumpkin the farthest. Held on the first weekend in November every year since, the event has grown from three teams and a small group of friends competing to sixty teams with more than thirty thousand spectators. The longest launch that first year was 126 feet. The winning launch in 2002 went nearly 3,900 feet!

Contestants have come up with many ingenious designs for their launch vehicles. Most are catapult-like machines that use springs, power bands, hydraulic lines, or compressed air. The event has only four rules:

1 - pumpkins must weigh between eight and ten pounds

2 - pumpkins must be in one piece when launched

3 - no part of the launching machine may cross the starting line, and

4 - no explosives allowed!

▶ Some Punkin' Chunkin' catapults resemble the three-dimensional figure shown below. How many faces, edges, and vertices does this solid have?

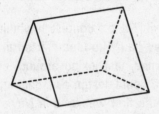

Know It All Approach

Use the **Know It All Approach** to find your answer. Start by reading the question carefully. Make note of the words and information that you need in order to answer the question. The fact that the solid resembles some Punkin' Chunkin' catapults is irrelevant. Your task is to count faces, edges, and vertices. All the information you need is contained in the diagram of the solid.

Next, answer the question. Examine the diagram carefully. The solid has two triangular faces and three rectangular faces. Hence, the total number of faces equals five. The base has four edges, the sides have two edges each, and the top has one edge. The total number of edges equals nine. The base has four vertices (corners) and the top has two. Therefore, the total number of vertices equals six.

Don't forget to check your answer. A good way to double-check your answer is to use the formula.

$$F + V - E = 2$$
$$5 + 6 - 9 = 2$$
$$11 - 9 = 2$$
$$2 = 2$$

Finally, write your answer in the space provided:

Number of Faces = 5, Number of Edges = 9, Number of Vertices = 6

Directions: Read the passage below and answer the questions that follow.

Zip It!

In 1851, Elias Howe, inventor of the sewing machine, applied for a patent for an "Automatic, Continuous Clothing Closure." For reasons unknown, he did not attempt to market his new gadget. Thirty-nine years later, Whitcomb Judson, an American engineer from Chicago, Illinois, invented a hook-and-eye shoe fastener with locking teeth that was quite similar to Howe's device. He patented his new "Clasp Locker" in 1893, and later that same year introduced it at the Chicago World's Fair. Despite the fact that his shoe fastener was awkward to use, Judson is recognized as the father of the zipper. However, he was never able to successfully market his new invention.

Gideon Sundbach, a Swedish-American engineer, improved on Judson's inelegant design.

In 1923, the B.F. Goodrich Company came up with the name "zipper." Gradually, manufacturers started putting zippers in clothing. In the late 1940s, zippers became trendy as a result of the promotion by the fashion industry.

1. In parallelogram *QRST* shown below, $\overline{QR} = 5x - 1$, $\overline{RS} = 3x$, and $\overline{ST} = 4x + 2$. What is the length of $\overline{QT}$?

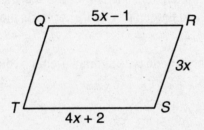

2. What is the value of *x* in the figure shown below?

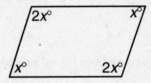

A   30
B   36
C   60
D   72

3. The solid shown below would be classified as a

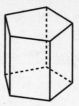

A   cone
B   prism
C   pyramid
D   tetrahedron

Directions: Read the passage below and answer the questions that follow.

The Uncertain Sultan

On August 27, 1896, a small fleet of British warships began bombarding Zanzibar, a small island off the east coast of Africa. The shelling started at 9:00 A.M. Forty-five minutes later, Zanzibar surrendered. It was the shortest war in recorded history. For many years, Great Britain and Germany had been involved in a political squabble over which country should govern the African island nation of Zanzibar.

By the early 1890s, the British had discreetly assumed control. However, when the sultan of Zanzibar died in 1896, his son assumed the throne and proclaimed the island's independence. Fearing that Germany would eventually take over the island, the British issued the new sultan an ultimatum—resign or be deposed by force. The young monarch declined the offer. To the contrary, he amassed an army of 2,500 troops to defend his newly independent state. When the British ships started their artillery salvo, the new sultan had second thoughts. After all, the only cannon available for Zanzibar's defense had not been fired since 1658.

At 9:45 A.M., the new sultan surrendered after obtaining a guarantee of safe haven from the Germans.

4. Pretend the figure below represents the walls of the sultan's palace. In figure *WXYZ*, $\overline{WX} = 4x - 2$, $\overline{XY} = 3x + 3$, $\overline{YZ} = 2x + 8$, and $\overline{WZ} = x + 13$. Prove that figure *WXYZ* is a rhombus. **Show all work.**

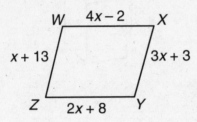

5. What is the value of *x* in the figure shown below?

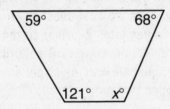

A 88
B 102
C 112
D 122

6. What is the measure in degrees of each interior angle of the regular polygon shown below? **Show your work.**

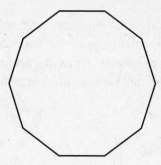

_____

Subject Review

In Chapter 11, you learned that a polygon is a multi-sided closed figure. You were shown how to find the measure of the angles of a regular polygon. You learned the properties of different quadrilaterals. You were introduced to different types of solid figures. Lastly, you learned how to calculate the number of faces, edges, and vertices of various three-dimensional shapes.

Plus, you know these strange-but-true facts.

What is "Punkin' Chunkin'"?

Punkin' Chunkin' is an annual contest in which teams attempt to launch a pumpkin as far as they can. Recent winners have launched pumpkins about 4,000 feet!

Who invented the zipper?

Whitcomb Judson of Chicago, IL, gets credit for the useful invention. However, a few people contributed their designs to achieve the product we know and love today.

What was the shortest war in world history?

Great Britain fought Zanzibar on August 27, 1896. The whole war lasted only 45 minutes!

Directions: Read the passage below and answer the questions that follow.

King-Size Cookie

Joker

On May 17, 2003, Immaculate Baking Co. baked the World's Largest Cookie! The cookie was approximately one hundred feet in diameter, which is slightly longer than the length of a basketball court. It weighed more than forty thousand pounds—about as much as four average-size elephants! It was baked in a structureless oven with a base made out of sheet aluminum. The cookie dough was frozen in advance, and then cut into one-foot squares about $\frac{1}{2}$ inch thick. With the oven heated to a temperature of 350° F (177° C), it took almost six hours to bake the giant chocolate-chip cookie. It contained more than thirteen million semisweet chocolate chunks weighing six thousand pounds.

1. A baker has x pounds of semisweet chocolate chips. He uses twenty-four pounds of the chips baking cookies, and then receives a new supply of $5y$ pounds. Which of the following expressions represents the weight of chocolate chips he has now?

    A $x + 24 + 5y$
    B $x + 24 - 5y$
    C $x - 24 + 5y$
    D $x - 24 - 5y$

2. Magic Bagel Cafe offers two specials. The first special includes a dozen bagels and one pound of cream cheese for $8.69. The second special comes with twenty-five bagels and three pounds of cream cheese for $21.12. These specials are shown in the system of equations below, where b represents the cost of one bagel and c represents the cost of one pound of cream cheese.

    $12b + c = \$8.69$

    $25b + 3c = \$21.12$

Solve the system of equations to determine the cost of one pound of cream cheese. **Show all work.**

3. The line on the graph below represents an equation. What is its slope?

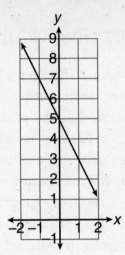

A -2

B $-\dfrac{1}{2}$

C $\dfrac{1}{2}$

D 2

4. Factor: $x^2 - 5x - 14$

A $(x + 7)(x + 2)$

B $(x + 7)(x - 2)$

C $(x - 7)(x + 2)$

D $(x - 7)(x - 2)$

5. Which of the following is the graph of the equation $y = 2x + 3$?

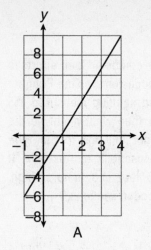

A

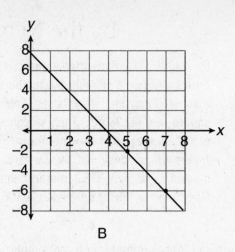

B

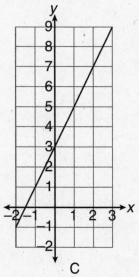

C

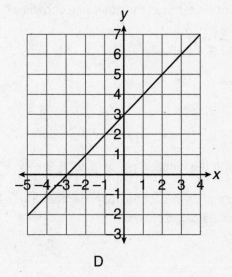

D

Directions: Read the passage below and answer the questions that follow.

By the Numbers

On July 29, 2001, more than a thousand people combined their efforts to create a gigantic painting-by-numbers. The occasion was the Peugeot Summer Party in Hertfordshire, England. The painting measured 455 square feet (42.26 m²). This is slightly larger than what might be the world's largest mural-by-numbers. That honor belongs to the *Pine Rivers Mural*, designed by Janet Skinner-Moschella of Australia. It measures 52 feet (15.8 meters) long by 6 feet (1.8 meters) high. Completed in 1998, it contained 20,096 color-coded sections.

6. The formula for the perimeter of a rectangular painting is $P = 2l + 2w$. What is the value of l in terms of the other two variables?

A $l = P - 2w$

B $l = P + 2w$

C $l = \dfrac{(P + 2w)}{2}$

D $l = \dfrac{(P - 2w)}{2}$

7. Consider the function $y = x^2 + 6x + 8$.

Make a table of values for the function from $x = -2$ to $x = 2$, and then graph the equation on the coordinate grid provided.

x	$y = x^2 + 6x + 8$	y
−2		
−1		
0		
1		
2		

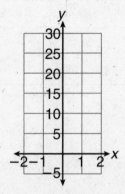

Chapter 12: Area and Perimeter

Who is known as the "petite Picasso"?

What were some of the world's largest cheeses ever created?

How big is the world's largest kite?

Area

Area is the number of square units needed to cover a surface enclosed by a geometric figure. Imagine that your kitchen measures fifteen feet by twelve feet. Think of the kitchen floor pictured below as a surface enclosed by a geometric figure, namely, a rectangle. If you wanted to cover the floor with ceramic tiles, and each tile is a one-foot square, how many tiles would you need?

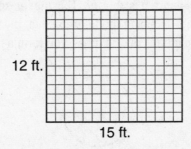

12 ft.

15 ft.

You could count all the spaces and figure out that you would need 180 tiles. However, there is an easier way. The formula for the area of a rectangle is $A = lw$, where A is the area in square units, l is the length, and w is the width.

$A = lw$

$A = 15 \cdot 12$

$A = 180$ sq. ft.

Each of the common geometric figures has its own formula for area. For example, a square is a special rectangle all of whose sides are congruent. Therefore, to find the area of a square, you could change $A = lw$ to $A = s \cdot s$ or $A = s^2$, in which s is the measure of one side of the square.

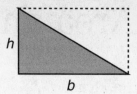

The formula for the area of a triangle is $A = \frac{1}{2}\,bh$, in which b is the base of the triangle and h is the height. This makes sense if you think of a triangle as one half of a rectangle, as shown in the figure above. The area of the rectangle would be $A = bh$. Since each triangle covers exactly one half of the rectangle, the area of each triangle would be $A = \frac{1}{2}\,bh$.

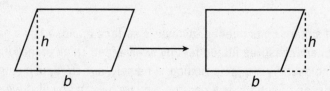

The formula for the area of a parallelogram is $A = bh$. This makes sense if you think of a parallelogram as a slanted rectangle. Examine the figure above. If you cut a right triangle from one side of the parallelogram and paste it back on the other, you form a rectangle whose area is $A = bh$.

Next, you need to consider a geometric figure that does not have any sides (or perhaps it has an infinite number of sides)—the circle. Ancient civilizations, from Babylonia, Egypt, and Greece, discovered the relationship known today as π (pi). The formula for the area of a circle is $A = \pi r^2$, where π equals 3.14 or $\frac{22}{7}$, and r is the radius of the circle. Try the sample question on the next page.

▶ What is the area of a circle with a diameter of 14 inches?

This one's a little tricky! Notice that the question does not give you the radius. Of course, you should know that the radius is equal to one half the diameter. Hence, the radius is 7 inches.

$$A = \pi r^2 \qquad\qquad A = \pi r^2$$

$$A = 3.14 \cdot 7^2 \qquad\qquad A = \frac{22}{7} \cdot 7^2$$

$$A = 3.14 \cdot 49 \qquad\qquad A = \frac{22}{7} \cdot 49$$

$$A = 153.86 \text{ sq. in.} \qquad A = 154 \text{ sq. in.}$$

As you can see, the answer is slightly different when you use $\pi = \frac{22}{7}$. Be careful whenever you answer a question about circles. Although 3.14 is considered the more accurate value for π, some questions tell you which value you **must** use.

Test questions often ask you to find the area of a complex figure, that is, a combination of two or more common geometric shapes. Look at the sample question below.

▶ What is the area of the figure shown below?

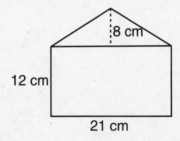

The first step is to realize that the figure shown above is a combination of a rectangle and a triangle. You must find the area of each smaller figure, and then add them.

$$A = lw \qquad\qquad A = \frac{1}{2} bh$$

$$A = 21 \cdot 12 \qquad\qquad A = \frac{1}{2} \cdot 21 \cdot 8$$

$$A = 252 \text{ cm}^2 \qquad\qquad A = 84 \text{ cm}^2$$

The total area of the figure is $252 + 84 = 336 \text{ cm}^2$.

Perimeter

Perimeter is the distance around a geometric figure. You can find the perimeter of any polygon by simply adding all its sides. Check out the sample question below.

▶ Robin's father built a U-shaped deck (see figure below) around the patio. What is the perimeter of the deck?

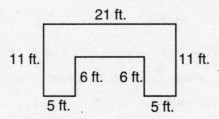

A   60 ft.
B   65 ft.
C   71 ft.
D   76 ft.

To find the perimeter of the deck, you need to add all the sides. This should be easy, but be careful! There is one side that is not labeled. To figure out the length of the missing side (bottom-middle), you have to subtract the two bottom sides (5 feet each) from the top side (21 feet). The missing side must be 11 feet long.

$$P = 21 + 11 + 5 + 6 + 11 + 6 + 5 + 11 = 76 \text{ ft.}$$

The perimeter of the deck is 76 ft. Therefore, answer choice (D) is correct. Notice that the answer is in feet, not square feet. This is a linear distance and has nothing to do with the area needed to cover a surface.

Each of the common geometric figures has its own formula for perimeter, as shown in the chart below. Notice that the perimeter of a circle is called its "circumference." This is to distinguish it from perimeter, which is usually a straight-line distance. "Circumference" literally means "to carry around."

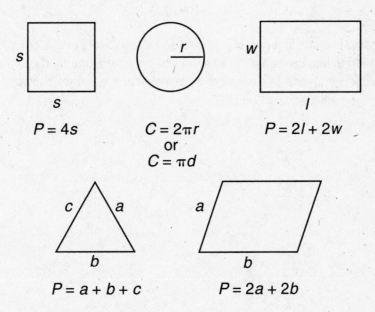

Just as with area, test questions sometimes ask you to find the perimeter of a complex figure. Examine the sample question below.

▶ Mrs. Graham is having a new window installed. It is round on top and square on the bottom, as shown in the figure below. What is the perimeter of the window?

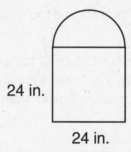

24 in.

24 in.

Obviously, the window is a combination of a square and a semicircle (half a circle). You might suspect that you need to find the perimeter of each smaller figure, and then add them. Since the diameter of the circle is the same line segment as one side of the square, the diameter equals 24 inches. Therefore, the radius equals 12 inches.

$P = 4s$ $C = 2\pi r$

$P = 4 \cdot 24$ $C = 2 \cdot 3.14 \cdot 12$

$P = 96$ in. $C = 75.36$ in.

Do these results mean that the perimeter of the window is 96 + 75.36, or 171.36 inches? Not so fast! For area problems, the surface covered by one figure does not affect the surface covered by an adjoining figure. However, this is not the case for perimeter problems.

Remember, perimeter is the distance **around** a geometric figure. Only **three** sides of the square go around the window. Only these three sides make up part of the window's perimeter. The fourth side is the diameter of the semicircle. It is **inside** the window; hence, it is **not** part of the perimeter. The remainder of the window's perimeter is the semicircle. The distance around the semicircle is one half the total circumference. Therefore, the perimeter of the window is:

$$P = 3 \cdot 24 + \frac{1}{2} \cdot 75.36$$

$$P = 72 + 37.68$$

$$P = 109.36 \text{ in.}$$

Scale Factors vs. Area and Perimeter

There is one more concept you need to consider. Imagine that you doubled all the measurements in a figure. How would it affect the area and the perimeter? Look at the sample problem below.

▶ Dustin is making sails for a model sailboat. He has to double the length of all sides of the sails below. By what factor will the area of the sails increase? By what factor will the perimeter of the sails increase? The shapes of the sails are shown below.

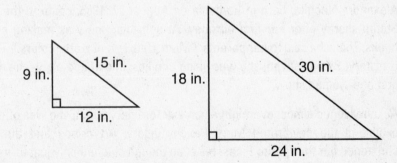

You need to find the area and perimeter of each sail, and then compare them.

Small Triangle

$A = \frac{1}{2}bh$

$A = \frac{1}{2} \cdot 12 \cdot 9$

$A = 54$ sq. in.

$P = a + b + c$

$P = 9 + 12 + 15$

$P = 36$ in.

Large Triangle

$A = \frac{1}{2}bh$

$A = \frac{1}{2} \cdot 24 \cdot 18$

$A = 216$ sq. in.

$P = a + b + c$

$P = 18 + 24 + 30$

$P = 72$ in.

By doubling the length of all the sides, the area increased by a factor of four. However, the perimeter increased by a factor of only two. The reason for this should be apparent. To find perimeter, a one-dimensional property, you **add** the sides. To find area, a two-dimensional property, you **multiply** the sides. Picture in your mind a square that measures 2 feet by 2 feet. Its perimeter is 8 feet, and its area is 4 square feet. Now, double the dimensions. The new square is 4 feet by 4 feet. Its perimeter is 16 feet, and its area is 16 square feet. The perimeter increased by a factor of 2, but the area increased by a factor of 2^2, or 4.

Directions: Read the passage below and answer the question that follows.

The Petite Picasso

Alexandra Nechita, born in Romania on August 27, 1985, came to the United States shortly after her first birthday. At age four, she was making coloring books. Then she said to her parents, "Mom, Dad, I want water colors." The rest is history! By age seven, she was using acrylics, and at age eight, she had her first one-woman show.

Acknowledged almost overnight as an extraordinary talent, she was offered an exhibit at the celebrated Mary Paxon Gallery. Art critics and the media christened her the "petite Picasso," even though she had only just turned nine years old. She is an abstract painter in the cubist style. She counts Picasso and Klee among her inspirations. Today, as a young adult, she is among the world's most recognized artists. She has produced more than three hundred paintings, many of which have sold for more than a million dollars. She lives with her family in Los Angeles, California.

▶ "The Wine Taster," is a painting by Alexandra Nechita. It measures $41\frac{1}{2}$ inches by $26\frac{3}{4}$ inches. What is the area of this painting?

_____

Know It All Approach

First, read the question carefully. Note the words and information that you need in order to answer the question. The important information includes the length $\left(41\frac{1}{2} \text{ inches}\right)$ and the width $\left(26\frac{3}{4} \text{ inches}\right)$.

Next, answer the question. Use the formula for the area of a rectangle: $A = lw$. If you have access to a calculator, you might want to convert the fractions to decimals.

$$A = lw \qquad\qquad A = lw$$

$$A = 41\frac{1}{2} \cdot 26\frac{3}{4} \qquad A = 41.5 \cdot 26.75$$

$$A = 1,110\frac{1}{8} \text{ in.}^2 \qquad A = 1,110.125 \text{ in.}^2$$

After you have found your answer, double-check to make sure you've made no mistakes in your calculations. Because you multiplied to get your answer, a good way to double-check is to divide your answer by either the length or the width.

$$1,110.125 \div 41.5 = 26.75$$

Finally, write your answer in the space provided. Make sure you label your answer with the proper unit. The answer to this question needs to be in square inches.

Directions: Read the passage below and answer the questions that follow.

The Big Cheese

In 1866, a Canadian cheese factory created a cheese that weighed 7,300 pounds. It measured 3 feet high and 7 feet in diameter. One of the largest cheeses ever made was a 22,000 pounds, 11-ton, round cheese with a circumference of 28 feet (8.5 meters). To make it, they used the equivalent of a day's milk from ten thousand cows. However, the world record for the largest cheese was set by the Agropur Cooperative Agro-Alimentaire, a leader in the

Canadian dairy industry. The company is owned by almost five thousand dairy producers. In 1995, it made a cheddar cheese weighing 57,508 pounds for Loblaws Supermarkets Ltd. of Quebec, Canada. More than 245,000 kilograms (540,000 pounds) of milk were used in making the cheese.

1. The 22,000-pound round cheese mentioned on the previous page has an actual circumference of 28.26 feet. What is the area of its uppermost surface? (**Use $\pi = 3.14$**)

   A   14.13 ft.^{2}
   B   63.585 ft.^{2}
   C   254.34 ft.^{2}
   D   626.92 ft.^{2}

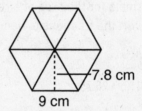

9 cm   7.8 cm

2. The diagram above shows a regular hexagon. It is composed of 6 congruent triangles. Each triangle has a base of 9 centimeters and a height of 7.8 centimeters. What is the area of the entire hexagon?

   A   35.1 cm^2
   B   70.2 cm^2
   C   210.6 cm^2
   D   421.2 cm^2

3. A square room has a perimeter of forty-eight feet. How many square feet of carpeting are needed to cover the floor with wall-to-wall carpeting?

   A   48 ft.^{2}

   B   144 ft.^{2}

   C   576 ft.^{2}

   D   2,304 ft.^{2}

4. As shown in the diagram below, the school's running track is in the shape of a rectangle with $\frac{1}{2}$ circle on each end. The longest running event at a track meet requires the participants to complete $4\frac{3}{4}$ laps.

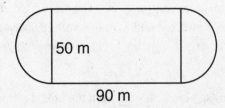

50 m

90 m

What distance in meters must each participant run? **Show all work.**

Directions: Read the passage below and answer the questions that follow.

High-Flying Giant

The world's largest kite in the world that has actually been flown is called the "Megabite." Peter Lynn of New Zealand designed it. It covers an area of 680 square meters (6,885.9 square feet) when laid flat. It is 64 meters (210 feet) long (including a 30-foot tail) and 22 meters (72 feet) wide. On September 7, 1997, at the Bristol International Kite Festival in the United Kingdom, the kite flew for twenty-two minutes and fifty-seven seconds. Incredibly, the Megabite is flown off a single line, though there are two additional steering lines attached to help control it in extreme winds. The kite, which took approximately 350 hours to build, is made from 20,000 yards of fabric. To make it easier to transport, it can be broken down into a number of zippered sections.

The Megabite doesn't look very big from the outside. However, you can really get a feel for its size if you go **inside** it. In point of fact, a few curious spectators took advantage of the opportunity to actually fly their own small kites inside the Megabite!

The "Megabite," the world's largest kite

5. Without its tail, the Megabite covers an area of 5,940 square feet when laid flat. If another kite with a triangular shape had the same area with a base of 72 feet, what is its height without the tail?

   A   41.25 ft.
   B   82.5 ft.
   C   165 ft.
   D   210 ft.

6. At a medieval-style park, actors dressed as knights, bishops, kings, queens, and pawns and assumed the roles of the various chess pieces on a giant chessboard. The board is 40 feet by 40 feet square. If the board is made up of 64 identical smaller squares, what is the area of each smaller square?

   A   25 ft.^{2}
   B   40 ft.^{2}
   C   1,600 ft.^{2}
   D   2,560 ft.^{2}

7. In the diagram below, a circle is inscribed in a square. Each side of the square measures 20 inches.

20 in.

What is the area of the circle?

   A   62.8 sq. in.
   B   125.6 sq. in.
   C   314 sq. in.
   D   1,256 sq. in.

Subject Review

In Chapter 12, you learned that area is the number of square units needed to cover a surface enclosed by a geometric figure. You also learned that perimeter is the distance around a geometric figure. You discovered that the common geometric figures each have their own formula for area and perimeter. You learned how to find the area and perimeter of complex geometric figures. Lastly, you found out how changes in the dimensions of a figure affect area and perimeter.

As an added bonus, you now know the following facts.

Who is known as the "petite Picasso"?
Alexandra Nechita has earned the nickname because of her incredible artistic talent at such a young age.

What were some of the world's largest cheeses ever created?
A block of cheese in the 1800s weighed 7,300 pounds. It was so popular that it inspired poetry. More recently, an immense block of cheddar cheese was created in Quebec, Canada, that weighed more than 57,000 pounds!

How big is the world's largest kite?
The Megabite covers 680 square meters (6,885.9 sq. ft.) of area when laid flat. As of 2003, it's the biggest kite to have ever flown. Maybe someday a bigger kite will fly the skies.

Chapter 13: Surface Area and Volume

How many people can eat from one enormous chocolate bar?

What will make shopping at the supermarket less tedious in the future?

What's the easiest way to rewire a school for the Internet?

Surface Area

Have you ever wrapped a gift that came in a box? If you have, then you know you had to figure out how much wrapping paper was needed to cover all the surfaces of the box. In other words, you had to figure out the surface area. **Surface area** is the sum of the areas of all the faces or surfaces of a solid. Look at the sample question below.

▶ What is the total surface area of the cardboard box shown below?

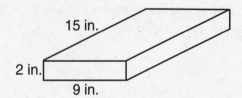

A   52 in.^{2}
B   183 in.^{2}
C   270 in.^{2}
D   366 in.^{2}

Like all rectangular prisms, a box has six faces. Each of these six rectangular faces has its own area. It may be easier to think of them as **three pairs** of faces. In this particular box, each of the two small faces measures 2 in. by 9 in. Each of the two large faces measures 9 inches by 15 inches, and each of the two remaining (midsize) faces measures 2 inches by 15 inches.

The total surface area is the sum of the areas of all six faces.

Small Face	Large Face	Midsize Face
$A = lw$	$A = lw$	$A = lw$
$A = 2 \cdot 9$	$A = 9 \cdot 15$	$A = 2 \cdot 15$
$A = 18$ in.^{2}	$A = 135$ in.^{2}	$A = 30$ in.^{2}

Total surface area $= (2 \cdot 18) + (2 \cdot 135) + (2 \cdot 30)$

$36 + 270 + 60$

366 in.^{2}

The total surface area of the box is 366 square inches. Therefore, answer choice (D) is correct. Because many common objects are rectangular prisms (e.g., boxes, blocks, cubes, etc.), mathematicians have come up with a formula for their surface areas. Try solving the problem above using the formula given below.

$SA = 2lh + 2lw + 2wh$

$SA = (2 \cdot 15 \cdot 2) + (2 \cdot 15 \cdot 9) + (2 \cdot 9 \cdot 2)$

$SA = 60 + 270 + 36$

$SA = 366$ in.^{2}

As you can see, the answer is the same. Therefore, the formula must be valid. Of course, not all containers are boxes. Try the next sample question.

► For his birthday, Anita bought her brother a container of three tennis balls, as shown on the next page. She wants to gift-wrap the container, which has a diameter of 3.2 inches and a height of 10 inches. What is the total surface area of the container?

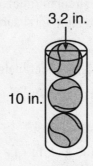

3.2 in.

10 in.

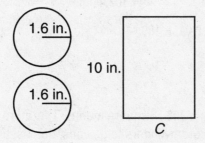

1.6 in.

1.6 in.

10 in.

C

A   105.50 in.²
B   108.52 in.²
C   110.53 in.²
D   116.56 in.²

You need to ask yourself, "What makes up the surface area of a cylinder?" The top and the bottom are congruent circles ($A = \pi r^2$). That part is easy. What about the middle? Try to imagine carefully peeling off the label from a can of soup. What is its shape? It's a rectangle!

The length of the rectangle is equal to the circumference ($C = 2\pi r$) of either circle. The width is equal to the height of the cylinder. Therefore, the total surface area of a cylinder is the sum of the areas of the two circles and the rectangle.

Top or Bottom Circle	Middle Rectangle
$A = \pi r^2$	$A = Ch$
$A = 3.14 \cdot 1.6^2$	$A = 2\pi rh$
$A = 3.14 \cdot 2.56$	$A = 2 \cdot 3.14 \cdot 1.6 \cdot 10$
$A = 8.0384$ or 8.04 in.^{2}	$A = 100.48$ in.^{2}

Total surface area $= 2 \cdot 8.04 + 100.48$

16.08 + 100.48

116.56 in.^{2}

The total surface area of the box is 116.56 square inches. Therefore, answer choice (D) is correct. Many objects you see every day are cylinders (e.g., cans, bottles, tubes, etc.). Consequently, mathematicians have come up with a formula for their surface areas, too. Try solving the problem above using the formula given below.

$SA = 2\pi r^2 + 2\pi rh$

$SA = (2 \cdot 3.14 \cdot 1.6^2) + 2 \cdot 3.14 \cdot 1.6 \cdot 10$

$SA = (6.28 \cdot 2.56) + 100.48$

$SA = 16.08 + 100.48$

$SA = 116.56$ in.^{2}

Again, the answer is the same. Therefore, this formula must be valid as well.

Know It All! High School Math

Volume

Earlier, you learned about the outside of a box or its surface area. Now, you will learn about the inside! After all, what do people do with a box? They put things in it! The amount of stuff you can put into a box depends on how much space is inside the box. **Volume** is the number of cubic units needed to fill the space occupied by a solid. Check out the sample question on the next page.

▶ Miguel's little sister is always leaving her toy blocks all over the floor. Miguel bought his sister a box (shown below) to store the toy blocks in when she is not playing with them. If each block is a 5-centimeter cube, how many blocks can be stored inside the box?

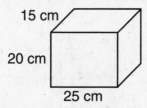

15 cm

20 cm

25 cm

A   12
B   30
C   60
D   150

The first step is to find the volume of the box. Then, you need to find the volume of each toy block. Finally, you must divide to figure out how many blocks can fit inside the box. The formula for the volume of a box (or any rectangular prism) is $V = lwh$.

Box

$V = lwh$

$V = 25 \cdot 15 \cdot 20$

$V = 7{,}500 \text{ cm}^3$

Toy Block

$V = lwh$

$V = 5 \cdot 5 \cdot 5$

$V = 125 \text{ cm}^3$

If the volume of the box is 7,500 cubic centimeters, and each toy block occupies 125 cubic centimeters of space, then the number of blocks that can fit inside the box is 7,500 ÷ 125, or 60. Sixty blocks can be stored in the box. Therefore, answer choice (C) is correct.

Take a moment to think about some of the liquids you use almost every day. They might include drinking water, soda, juice, milk, and shampoo, just to name a few. In what type of container are these liquids usually sold? You guessed it—cylinders!

The formula for the volume of a cylinder is $V = \pi r^2 h$. Try the sample question below.

▶ The diagram below shows a tall lemonade glass with a radius of 1.5 in. and a height of 6 inches. What is the volume, in cubic centimeters, of the glass?

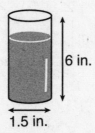

6 in.

1.5 in.

A   9.00 in.³
B   28.26 in.³
C   42.39 in.³
D   169.56 in.³

To find the volume of the glass, simply plug the radius and the height into the formula.

$$V = \pi r^2 h$$

$$V = 3.14 \cdot 1.5^2 \cdot 6$$

$$V = 3.14 \cdot 2.25 \cdot 6$$

$$V = 42.39 \text{ in.}^3$$

Answer choice (C) is correct. Of course, glasses come in many shapes. Some are even cone-shaped.

The formula for the volume of a cone is $V = \frac{1}{3} \pi r^2 h$.

▶ Brianna and her mother went to an old-fashioned ice-cream parlor. They were served milk shakes in a cone-shaped cup with a radius of 6 centimeters and a height of 14 centimeters. What was the volume, in cubic centimeters, of the cup?

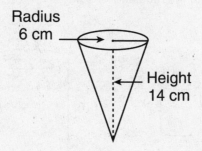

Radius
6 cm

Height
14 cm

A   474.77 cm³
B   527.52 cm³
C   1,055.04 cm³
D   1,582.56 cm³

Plug the radius and the height into the formula for the volume of a cone.

$$V = \frac{1}{3} \pi\ r^2 h$$

$$V = \frac{1}{3} \cdot 3.14 \cdot 6^2 \cdot 14$$

$$V = \frac{1}{3} \cdot 3.14 \cdot 36 \cdot 14$$

$$V = 527.52 \text{ cm}^3$$

The volume of the paper cup is 527.52 cubic centimeters. Answer choice (B) is correct.

What effect does doubling the dimensions of a solid figure have on volume? Try the sample problem below and find out.

▶ A company that sells peaches in a can wants to market a super-size can that has twice the diameter and twice the height of the regular can. What is the ratio of the volume of the new super-size can to the volume of the regular-size can?

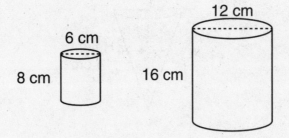

A   2 to 1
B   4 to 1
C   8 to 1
D   16 to 1

To find the requested ratio, you need to calculate the volume of each cylinder, and then divide. Use the formula $V = \pi r^2 h$. Be sure to divide the diameter by 2 to find the radius.

Cylinder A	Cylinder B
$V = \pi r^2 h$	$V = \pi r^2 h$
$V = 3.14 \cdot 3^2 \cdot 8$	$V = 3.14 \cdot 6^2 \cdot 16$
$V = 3.14 \cdot 9 \cdot 8$	$V = 3.14 \cdot 36 \cdot 16$
$V = 226.08 \text{ cm}^3$	$V = 1{,}808.64 \text{ cm}^3$

Ratio: $\dfrac{\text{Volume of Cylinder } B}{\text{Volume of Cylinder } A} = \dfrac{1{,}808.64}{226.08} = 8$

The requested ratio is $\frac{8}{1}$ or 8 to 1. Therefore, answer choice (C) is correct. Do you notice a pattern here? When you double the dimensions of a geometric figure, the perimeter increases by a factor of 2 (2^1), the area increases by a factor of 4 (2^2), and the volume increases by a factor of 8 (2^3). This makes a lot of sense when you consider that perimeter is a one-dimensional property, area is a two-dimensional property, and volume is a three-dimensional property.

Directions: Read the passage below and answer the question that follows.

Super-Sized Sweets

Joker

The record for the world's largest chocolate bar was set at the Euro-Chocolate 2000 Exhibition in Turin, Italy. Between March 16 and 19, 2000, Elah-Dufour United Food Companies Ltd. made an enormous *Novi* chocolate bar that was 124 inches long, 59 inches wide, and 17.7 inches thick (315 centimeters by 150 centimeters by 45 centimeters). It weighed an astounding 5,026 pounds (2,280 kilograms). It was the equivalent of 22,800 regular four-ounce chocolate bars.

The world's largest box of chocolates was assembled in New York City, New York, on February 14, 2000, in celebration of St. Valentine's Day. Measuring 15 feet by 15 feet by 2.5 feet (4.57 meters by 4.57 meters by 0.76 meters), the heart-shaped chocolate box contained 1,300 pounds (591 kilograms) of amaretto-filled chocolates.

► Imagine you constructed a rectangular box of chocolates that was 15 feet long, 15 feet wide, and 2.5 feet high. What would its volume be?

A 562.5 ft.^{3}

B 225.0 ft.^{3}

C 65.0 ft.^{3}

D 32.5 ft.^{3}

Know It All Approach

Start by reading the question carefully. You need to know the dimensions of the box: 15 feet long, 15 feet wide, and 2.5 feet high.

Once you've found all the info you need, answer the question. You need to plug into the formula the values given in the problem: $V = lwh$.

$$V = lwh$$
$$V = 15 \bullet 15 \bullet 2.5$$
$$V = 562.5 \text{ ft.}^3$$

Now check your answer. Because you multiplied to get your answer, a good way to double-check is to divide.

$$562.5 \div 15 = 37.5$$
$$37.5 \div 15 = 2.5$$

Because division resulted in the original dimensions, the answer is probably correct. Answer choice (A) is correct.

Directions: Read the passage below and answer the questions that follow.

Lost in the Supermarket?

Is there anything more frustrating than searching all over the supermarket for one item? This rite of passage may soon go the way of the dinosaur. Product designer Murray Laidlaw has developed a shopping cart that uses a global positioning system (GPS) to locate items on shelves. Thanks to communications satellite technology, it can direct shoppers to the appropriate aisle. A liquid crystal display screen resting atop the cart's handles displays arrows on a map of the store that indicate the locations of requested items. In fact, when they first walk in, shoppers can punch in the items they plan to purchase. The device will then lay out for them the shortest route to the items. It will also make available recipes and weekly store specials. According to Laidlaw, this device will make shopping trips less confusing and less time-consuming.

1. GPS systems use satellites to help track the movement of items on the ground. Imagine a satellite is 5 meters long, 1.8 meters wide, and 1.2 meters high. Its density is 125 kilograms per cubic meter. What would be the total mass of the satellite?

_____

2. A company that sells baking soda wants to design a box that will contain **four times** as much baking soda as their regular-size box represented in the figure below.

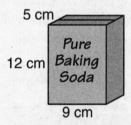

Which of the adjustments shown below will produce the appropriate box?

A   Double the width only.
B   Double the height only.
C   Double both the height and the width.
D   Double the length, the width, and the height.

3. A fish tank is 15 inches long and 8 inches wide. It can hold 1,320 cubic inches of water. What is the height of the fish tank?

A   5.5 in.

B   11 in.

C   57 in.

D   88 in.

Directions: Read the passage below and answer the questions that follow.

Cyber-Rat

Tie one end of a piece of string to a wire, and then tie the other end around Rattie's waist. That's all you have to do to get your school rewired. Rattie, a female albino rat, is helping California schools connect to the Internet. Dr. Judy Reavis, a medical doctor and head of her own computer company, trained Rattie to crawl behind walls and above ceilings while tied to a string that pulls computer wire. It took hours of work and a lot of patience to teach Rattie how to work her way through a maze of narrow passages. Dr. Reavis helps guide Rattie to the end by tapping on the walls and ceilings. Rattie is driven by the fact that she knows her favorite treats, cat food and gummi candies, will be waiting for her. Rattie has already helped rewire eight schools.

Rattie checks her work.

4. Computer and telephone wires are bundled neatly in a 75-centimeter length of plastic tubing with a radius of 2 centimeters. What is the volume of the tubing?

   A   235.5 cm³

   B   471 cm³

   C   942 cm³

   D   967.12 cm³

5. The swimming pool shown below is 1.3 meters high and can hold 13,200 gallons of water. A physics book states that one cubic meter is the same as 264 gallons. About how many meters is the radius of the swimming pool? **Show all work**.

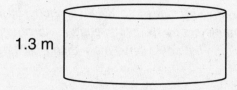

1.3 m

Subject Review

In Chapter 13, you learned that surface area is the sum of the areas of all the faces of a solid. You also were shown how to calculate the surface area of a rectangular prism and a cylinder. You learned that volume is the number of cubic units needed to fill the space occupied by a solid. You also were instructed on how to find the volume of a rectangular prism, a cylinder, and a cone. Lastly, you learned how doubling the dimensions of a solid affects the volume.

Furthermore, you're now hip to these cool facts.

How many people can eat from one enormous chocolate bar?
A single chocolate bar created in 2000 measured 124 inches long, 59 inches wide, and 17.7 inches thick (315 centimeters by 150 centimeters by 45 centimeters) and weighed 5,026 pounds (2,280 kg). That could serve more than 22,000 people each with a four-ounce serving.

What will make shopping at the supermarket less tedious in the future?
A GPS shopping cart is being developed that will tell you the location of any item in the store. Isn't that nice?

What's the easiest way to rewire a school for the Internet?
Have Rattie, the albino rat, do it for you!

Chapter 14: Similarity and Congruence

How big is the world's largest glass of orange juice?

How big is the world's largest wooden nickel?

How big is the world's largest tire?

Similarity

Which car would you rather have: the real car or the model? The picture below shows a model of Jeff Gordon's NASCAR racing car. Jeff Gordon's **actual** racing car is 110 inches long, 74 inches wide, and 51 inches high. The die-cast model of Jeff Gordon's racing car looks almost exactly like the real car, but it is 4.6 inches long, 3.1 inches wide, and 2.1 inches high. The engine in the real car has more than 750 horsepower. The model does not have an engine. The real car costs more than $1 million. The model costs about $80.

Not a difficult choice, is it? Almost anyone would rather have the real car. However, the real car and the model are **similar**. In mathematics, **similar figures** have the same shape but different sizes. The die-cast model is an exact replica of the actual car, built to a scale of 1:24. That means that each inch on the model represents 24 inches on the real car.

Similar figures have corresponding angles that are equal, and corresponding sides that are proportional. The sides of similar figures are related by a scale factor. If two figures are similar, they could match up exactly if one of them is magnified or shrunk.

Examine the two triangles in the figure below.

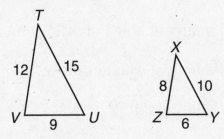

Is triangle *TUV* similar to triangle *XYZ*? If you measured them with a protractor, you would find that all the corresponding angles are congruent: $m\angle T = m\angle X$, $m\angle U = m\angle Y$, and $m\angle V = m\angle Z$. In addition, all the corresponding sides are proportional.

$$\frac{TU}{XY} = \frac{15}{10} = \frac{3}{2}$$

$$\frac{UV}{YZ} = \frac{9}{6} = \frac{3}{2}$$

$$\frac{TV}{XZ} = \frac{12}{8} = \frac{3}{2}$$

Therefore, $\triangle TUV$ is similar to $\triangle XYZ$. You can use this knowledge in the real world to find the length of an unknown side.

Michelle wanted to find out the height of a nearby lighthouse. She measured her shadow and the lighthouse's shadow one afternoon, which is shown in the diagram below.

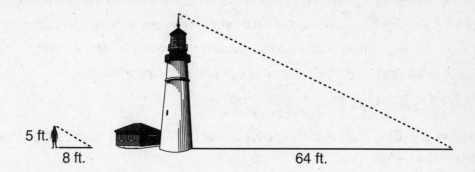

Michelle found that her shadow was 8 feet long at the same time the lighthouse's shadow was 64 feet long. Michelle is 5 feet tall. What is the height of the lighthouse?

A 40.0 ft.

B 62.5 ft.

C 64.0 ft.

D 102.4 ft.

Because Michelle measured her shadow and the lighthouse's shadow at the same time, the angle of the sun is the same for both. Also, both Michelle and the lighthouse stood on flat ground at right angles. Therefore, the corresponding angles in the two triangles are equal, and the two triangles are similar. You can set up a proportion to find the height, $h,$ of the lighthouse.

$$\frac{h}{5} = \frac{64}{8}$$

$$8h = 5 \cdot 64$$

$$\frac{8h}{8} = \frac{320}{8}$$

$$h = 40 \text{ feet}$$

The height of the lighthouse is 40 feet. Therefore, answer choice (A) is correct. You need to be careful when setting up proportions with similar figures. Each ratio must consist of a side of one figure over the **corresponding** side of the other figure. For example, in the above problem, if you had set up the proportion $\frac{h}{8} = \frac{64}{5}$, you would have gotten the wrong answer. That's because h, the unknown side in the larger triangle, and the 8-foot side in the smaller triangle are **not** corresponding sides. You need to be careful when you set up a proportion.

$$\frac{\text{height of lighthouse}}{\text{height of girl}} = \frac{\text{length of lighthouse's shadow}}{\text{length of girl's shadow}}$$

Some questions ask you to apply your knowledge of similar triangles to a coordinate grid. Look at the sample question below.

▶ Triangle ABC and triangle DEF are similar. The coordinates of the vertices of triangle ABC are shown on the graph below. If the coordinates of point E are $(-2, 4)$ and the coordinates of point F are $(-2, 1)$, which of the following could be the coordinates of point D?

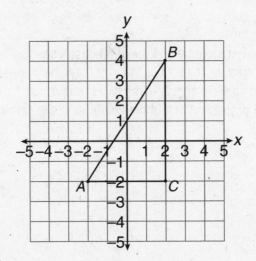

A $(-5, 1)$
B $(-2, 0)$
C $(-4, 1)$
D $(-4, 0)$

The first step is to plot points *E* and *F* on the graph. After you do, you should notice that $\overline{EF}$ is three units long, while its corresponding side, $\overline{BC}$, is six units long. So, you can assume that the length of the sides in △*DEF* is in a ratio of 1:2 to the length of the corresponding sides in △*ABC*. So, $\overline{DF}$ should be two units long (one-half the length of $\overline{AC}$, its corresponding side).

You know that the triangles are similar and that △ *ABC* is a right triangle, therefore, △*DEF* must also be a right triangle. If ∠*C* is a right angle, then ∠*F* must also be a right angle. For $\overline{DF}$ to form a right angle with $\overline{EF}$, point *D* must be located at either (−4, 1) or (0, 1). Consider the answer choices. (−4, 1) is among them, but (0, 1) is not. Therefore, the coordinates of point *D* are (−4, 1), and answer choice (C) is correct.

Congruence

The owner of a certain newsstand sells hundreds of newspapers every day. It so happens that he is blind. How does he know whether you are handing him a penny, a nickel, a dime, or a quarter? By the size! All coins have similar faces because the faces of all coins have a circular shape. (Coins are very thin cylinders. Their faces are always circles.) Coins have the same circular shape, but different sizes. All quarters have the same shape and the same size, therefore they are **congruent**. In mathematics, **congruent figures** have **the same shape and the same size. Congruent** shapes and **congruent** angles are identical.

Congruent figures have corresponding angles and corresponding sides that are identical. If two figures are congruent, and you place one on top of the other, they will match up exactly. Examine the two triangles in the figure below.

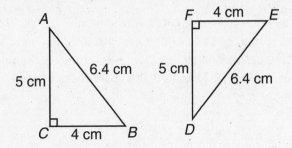

Is △ABC congruent to △DEF? If you measured each with a protractor, you would find that all the corresponding angles are congruent: m∠A = m∠D, m ∠B = m∠E, and m∠C = m∠F. In addition, all the corresponding sides are congruent. $\overline{AB} = \overline{DE}$, $\overline{BC} = \overline{EF}$, and $\overline{AC} = \overline{DF}$. If you cut out △DEF and placed it on top of △ABC, it would match up exactly, angle for angle and side for side. Therefore, the two triangles are congruent.

Unfortunately, in many cases you do not know all the measurements of each figure. Is there a way to determine if two triangles are congruent if some of the measurements are unknown? Actually, there are **four** ways!

Know It All! High School Math

Consider this question, which deals with four theorems you could use to determine the congruency of triangles.

▶ In the figure below, $\overline{AC} \cong \overline{BD}$ and $\overline{AD} \cong \overline{BC}$. Of the following, which theorem could you use to prove that $\triangle ACD \cong \triangle BDC$?

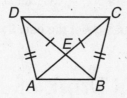

A AAS
B ASA
C SAS
D SSS

You're probably asking yourself, "What is this—some kind of code?" The letters in the answer choices stand for the four theorems you can use to prove two triangles congruent.

SSS (side-side-side)	3 sides of one triangle are congruent to their corresponding sides in the other triangle
SAS (side-angle-side)	2 sides and the included angle of one triangle are congruent to their corresponding sides and angles in the other triangle
ASA (angle-side-angle)	2 angles and the included side of one triangle are congruent to their corresponding angles and side in the other triangle
AAS (angle-angle-side)	2 angles and the non-included side of one triangle are congruent to their corresponding angles and side in the other triangle

Study the diagram carefully. Notice that △ACD and △BDC overlap. Also, notice the tick marks. Both $\overline{AC}$ and $\overline{BD}$ have one tick mark. This tells you they are congruent. Similarly, both $\overline{AD}$ and $\overline{BC}$ have two tick marks. They are congruent as well. What is the third side of each triangle? Trace the path from A to C to D, and then from B to D to C. $\overline{DC}$ is common to both triangles! It is the third side of both △ACD and △BDC. Obviously, $\overline{DC}$ is congruent to itself! Thus, three sides of △ACD are congruent to the three corresponding sides of △BDC. The two triangles are congruent according to the theorem SSS (side-side-side). Therefore, answer choice (D) is correct.

Directions: Read the passage below and answer the question that follows.

A Tall Drink of Juice

The world's largest glass of orange juice stood 8.5 feet tall and measured almost 5 feet in circumference. When filled, it weighed 4 tons (8,000 pounds) and contained more than 730 gallons of Florida Valencia orange juice. It was introduced by the Florida Department of Citrus in April 1998 to support National Minority Cancer Awareness Week. The American Cancer Society advocates a diet that is low in fat, high in fiber, and rich in vegetables and fruits—including orange juice. The glass was made of acrylic and took two weeks to build. It had a built-in refrigerator that maintained the temperature of the orange juice at 35 degrees. The orange juice was continuously circulated through the glass at a rate of 13 gallons per minute.

Stephanie was visiting Florida. After seeing the largest glass of orange juice, she decided to play golf. On one particular hole, golfers often hit their golf balls into the pond. To avoid doing this, Stephanie wants to find out how wide the pond is. She carefully measures the distance along line segments *AC*, *CE*, and *DE*, as shown in the diagram below. ∠*A* and ∠*E* are both right angles.

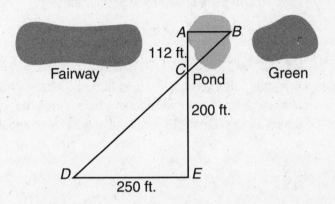

Part A   Explain in your own words why △*ABC* is similar to △*EDC*.

Part B   Write a proportion that you could use to find the distance from point *A* to point *B*.

Part C   Use the proportion from Part B to calculate the distance from point *A* to point *B* to the nearest foot. **Show all work**.

Know It All Approach

Begin by reading the question and looking over the diagram carefully. You want to make sure that you know about all the available information and that you understand what the question is asking.

Part A asks you to explain why △*ABC* is similar to △*EDC*. Similar triangles have corresponding angles that are equal and corresponding sides that are proportional. If you can show that all the angles in △*ABC* are equal to the corresponding angles in △*EDC*, then you will have proven that the triangles are similar.

$\angle A \cong \angle E$ They are both right angles.

$\angle BCA \cong \angle DCE$ They are vertical angles. All

vertical angles are congruent.

$\angle B \cong \angle D$ If two pairs of angles are congruent,

then the third pair is congruent.

Therefore, $\triangle ABC$ is similar to $\triangle EDC$ because their corresponding angles are congruent.

Part B asks you to write a proportion that you could use to find the distance from point A to point B. You can set up a proportion in which each ratio consists of a side of one triangle over the corresponding side of the other triangle. Since $\overline{AB}$ is the unknown side, it should appear first in the proportion.

$$\frac{\overline{AB}}{\overline{BE}} = \frac{\overline{AC}}{\overline{EC}} \qquad \frac{x}{250} = \frac{112}{200}$$

$\overline{AB}$ and $\overline{DE}$ are corresponding sides. $\overline{AC}$ and $\overline{EC}$ are corresponding sides.

Part C asks you to use the proportion from Part B to calculate the distance from point A to point B to the nearest foot.

$$\frac{x}{250} = \frac{112}{200}$$

$$200x = 250 \cdot 112$$

$$\frac{200x}{200} = \frac{28,000}{200}$$

$$x = 140 \text{ feet}$$

Once you have found the answer, double check it. A good way to do this is to set up a proportion with corresponding sides, and then see if both sides are equal.

$$\frac{\overline{AB}}{\overline{BE}} = \frac{\overline{AC}}{\overline{EC}}$$

$$\frac{140}{250} = \frac{112}{200} \qquad \text{Reduce to lowest terms.}$$

$$\frac{14}{25} = \frac{14}{25} \qquad \text{It checks. Your answer is correct.}$$

Now, make sure you have answered all parts of the question. You answered Part A by explaining why the two triangles are similar. You answered Part B by setting up a proportion to find the distance from point *A* to point *B*. And you answered Part C by calculating the distance from point *A* to point *B*. When writing your answer, be sure you write clearly.

Directions: Read the passage below and answer the questions that follow.

How Much Is a Wooden Nickel Worth?

A giant wooden nickel, most likely the world's largest, can be found in San Antonio, Texas. The giant wooden nickel is 13 feet, 4 inches in diameter and $5\frac{1}{2}$ inches thick. It is made of solid wood and weighs about 2,500 pounds. It contains enough wood to make about 400,000 standard-size nickels (out of wood). The design on the front side is dedicated to the Boy Scouts of America. The back side features a buffalo, similar to the back of the famous Buffalo nickel, minted in the United States from 1913 to 1938.

The history of the wooden nickel started in 1933 during the Great Depression. When their banks failed, the city of Blaine, Washington, issued round wooden coins. This marked the first appearance of wooden money in the United States.

1. The giant wooden nickel is similar to a real nickel. In the figure below, triangle *FGH* and triangle *JKL* are similar. Of the following statements, which is true about these two triangles?

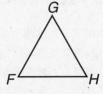

A   m ∠*F* = m ∠*G*
B   m ∠*F* = m ∠*H*
C   m ∠*F* = m ∠*J*
D   m ∠*F* = m ∠*L*

2. In the figure below, $\overline{QT} \cong \overline{ST}$, and $\angle QTR \cong \angle STR$. Which of the following theorems could you use to prove that $\triangle QRT \cong \triangle SRT$?

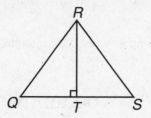

A AAS
B ASA
C SAS
D SSS

Directions: Read the passage below and answer the questions that follow.

Forget the Roller Coaster—I Want to Ride the Giant Tire

The world's largest tire is made of fiberglass and stands 86 feet tall. The Uniroyal-Goodrich Tire Company introduced it at the 1964–1965 World's Fair in New York. Attracting more than two million riders, it was an operating Ferris wheel with twenty-four barrel-shaped cars revolving around the center. At the conclusion of the Fair, Uniroyal transported the giant tire to Detroit as an advertising gimmick and a tourist attraction.

Michelin bought out Uniroyal—Goodrich in 1990. To promote its new self-sealing tire, Michelin revamped the giant tire, sticking a 10-foot nail into it. A new wheel cover reads, "Takes on Nails." The tire can be seen today from interstate-90 in Allen Park, Michigan.

3. The coordinates of the vertices of triangle *DEF* are shown on the graph below. Triangle *ABC* is similar to triangle *DEF*. If the coordinates of point *B* are (0, 4) and the coordinates of point *C* are (−2, 4), which of the following could be the coordinates of point *A*?

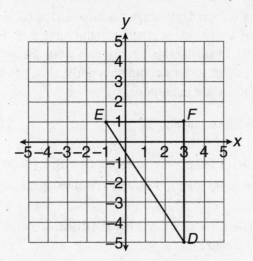

A   (−2, 1)
B   (−1, 1)
C   (1, −2)
D   (2, 1)

4. A 9-foot-high stop sign casts a shadow of 16 feet at the same time that a tree casts a shadow of 48 feet. What is the height of the tree?

9 ft.

16 ft.   48 ft.

Subject Review

In Chapter 14, you learned that similar figures are the same shape but different sizes. You also learned that congruent figures are the same shape and the same size. You were shown how to find the missing side in one of a pair of similar figures. You also were shown how to predict the coordinates of similar triangles plotted on a coordinate grid. Lastly, you learned four theorems that can be used to prove two triangles are congruent.

And you learned the following interesting tidbits.

How big is the world's largest glass of orange juice?
It is $8\frac{1}{2}$ feet tall and holds more than 730 gallons of orange juice.

How big is the world's largest wooden nickel?
It is 13 feet, 4 inches in diameter and $5\frac{1}{2}$ inches thick.

How big is the world's largest tire?
It is 86 feet tall (its diameter is 86 feet). You can visit it if you find yourself in Allen Park, Michigan.

Chapter 15: Angles and the Pythagorean Theorem

What is the angle of descent for the world's steepest roller coaster?

What is the record for the world's deepest submarine dive?

What is the world's record for the most time spent on a tightrope?

Angles

In a recent newspaper advertisement, the manufacturer of an LCD flat-panel TV/monitor claimed that the screen had a 160° viewing angle. What is a 160° angle?

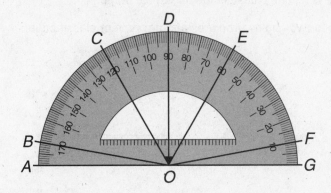

A circle measures 360°. The protractor pictured above is a half-circle, so it measures 180°. Thus, m∠AOG is 180°. What is m∠AOD? If you read the outer numbers on the protractor, you will see that as you move from point A to point O to point D, you generate an angle of 90°. Hence, m∠AOD is 90°.

What is m∠AOE? Again starting from point A, as you move to point O and then to point E, you generate an angle of 120°. Therefore, m∠AOE is 120°.

What is the purpose of the inner numbers on a protractor? Suppose you wanted to know m∠ *GOF*. If you used the outer numbers, you might think it is 170°. However, that is the wrong answer, because it is obvious that m∠ *GOF* is very small. When you measure an angle starting on the **right** side of the protractor (point *G*), you read the **inner** numbers. Thus, m∠ *GOF* is 10°. What is m∠ *BOF*? If you start at point *B*, and move to point *F*, you generate an angle of 160° (170 − 10). Hence, m∠ *BOF* is 160°.

The TV manufacturer is stating that you can watch this screen comfortably from anywhere inside the 160° angle.

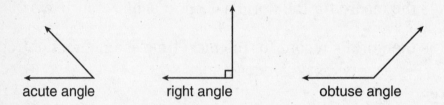

acute angle   right angle   obtuse angle

There are three basic types of angles, as shown in the diagram above. An **acute** angle measures less than 90°. A **right** angle measures exactly 90°. An **obtuse** angle measures more than 90°. Therefore, a 160° angle would be classified as an obtuse angle.

The relationship between two angles produces another type of classification.

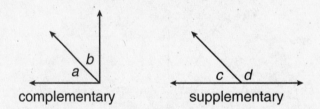

complementary   supplementary

When a line divides a right angle, the two angles formed are called **complementary** angles. Two angles are complementary if the sum of their measures is exactly 90°. In the diagram from the previous page, $\angle a$ and $\angle b$ are complementary.

When a line intersects a straight line, the two angles formed are called **supplementary** angles. Two angles are supplementary if the sum of their measures is exactly 180°. In the diagram from the previous page, $\angle c$ and $\angle d$ are supplementary. To check your understanding, examine the sample question below.

▶   In the figure below, what is m$\angle A$?

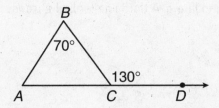

A   40°
B   50°
C   60°
D   130°

$\angle BCD$ is an external angle. It is formed by extending one side of $\triangle ABC$. The angles formed by the intersection of $\overline{BC}$ and $\overline{ACD}$ are supplementary. Thus, $\angle ACB$ is supplementary to $\angle BCD$. The sum of m$\angle ACB$ and m$\angle BCD$ is 180°. Hence, m$\angle ACB$ equals 180 − 130, or 50°.

Is that the correct answer? No, you need to find m$\angle A$! You should recall that the sum of the measure of all the angles of a triangle is 180°. You already know the measure of two of the angles in the triangle. Therefore, m$\angle A$ equals 180 − (70 + 50), or 180 − 120, which is 60°, so answer choice (C) is correct. Take a look at the sample question on the next page.

▶ The measure of an angle is 5x. Which of the following expressions represents the measure of its complement?

A $5x + 180$

B $5x + 90$

C $180 - 5x$

D $90 - 5x$

The sum of the measure of an angle and the measure of its complement is 90°. Thus, you need to find an angle whose measure, when added to 5x, equals 90. Since $5x + (90 - 5x) = 90$, answer choice (D) is correct.

When a line intersects two parallel lines, a completely new class of angles is formed. In the diagram below, line *f* is parallel to line *g*. A third line (called a transversal) intersects both line *f* and line *g*.

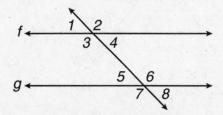

Angles that are side by side, that is, next to each other, are called **adjacent** angles. ∠1 and ∠2 are adjacent angles. ∠1 and ∠3 are also adjacent angles. All together in the diagram above there are **eight** pairs of adjacent angles! Can you name them?

When two parallel lines are cut by a transversal, all the adjacent angles formed are supplementary. For example, ∠5 and ∠6 are supplementary. ∠5 and ∠7 are also supplementary. Their sum is exactly 180°.

Angles that are opposite each other are called **vertical** angles. ∠1 and ∠4 are vertical angles. ∠2 and ∠3 are also vertical angles. All vertical angles are congruent.

Angles inside the two parallel lines are referred to as interior angles. Angles outside the two parallel lines are referred to as exterior angles. ∠3 and ∠6 are **alternate interior** angles. ∠4 and ∠5 are also alternate interior angles. Alternate interior angles are always congruent. ∠1 and ∠8 are **alternate exterior** angles. ∠2 and ∠7 are also alternate exterior angles. Alternate exterior angles are always congruent as well.

Angles in the same position relative to the two parallel lines and the transversal are called **corresponding** angles. ∠1 and ∠5 are corresponding angles. ∠3 and ∠7 are also corresponding angles. All corresponding angles are congruent. Look at the diagram below, then try the sample question on the next page.

▶ Given line *a* is parallel to line *b*

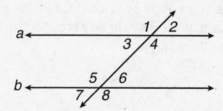

If m∠4 is 130°, what is the sum of m∠7 and m∠6?

    A   50°

    B   100°

    C   130°

    D   260°

If m ∠4 is 130°, then ∠8 is also 130° because they are complementary angles. Because ∠8 and ∠7 combine to form a straight line (line *b*), they must be supplementary. Hence, m∠7 = 180 − 130, or 50°. In addition, m∠6 is also 50° because ∠7 and ∠6 are vertical angles, and thus congruent to each other. So the sum of m∠7 and m ∠6 equals 50 + 50, or 100°. Answer choice (B) is correct. Try the next sample question.

▶ Given line *c* is parallel to line *d*, and m ∠1 = 65°, what is m∠2?

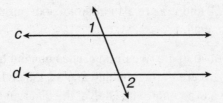

A   25°
B   65°
C   115°
D   165°

If m∠1 = 65°, then the measure of the angle adjacent to ∠1 (to its right) is 180 − 65, or 115°. In addition, that angle is congruent to ∠2 because they are alternate exterior angles. Therefore, m∠2 = 115°, and answer choice (C) is correct.

Pythagorean Theorem

Pythagoras, a famous Greek philosopher and mathematician, was born more than 2,500 years ago. The theorem you are about to discover bears his name. Although it was known to the Babylonians at least 1,000 years earlier, Pythagoras is credited with being the first to prove it mathematically. The Pythagorean theorem states, "In a right triangle, the square of the length of the hypotenuse is equal to the sum of the squares of the lengths of the legs." It may be easier to understand if you use a diagram.

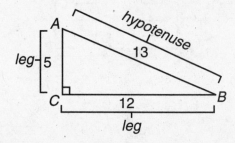

A **right triangle** is a triangle with one right angle. The two sides that form the right angle are the **legs**. The side opposite the right angle is the **hypotenuse.** The Pythagorean theorem is usually written as an equation: $c^2 = a^2 + b^2$, where c represents the length of the hypotenuse and a and b represent the lengths of the legs. See if it works for the triangle pictured in the diagram.

$$c^2 = a^2 + b^2$$

$$13^2 = 5^2 + 12^2$$

$$169 = 25 + 144$$

$$169 = 169$$

It works every time. Of course, a question will not always give you the length of all the sides. That is the whole purpose of the Pythagorean theorem. You can use it to determine the length of an unknown side. Look at the sample question on the next page.

Serena is painting the outside of her house. Her ladder extends to a length of 17 feet. She anchors the base of the ladder 8 feet from the side of the house. How far up the side of the house does the ladder reach?

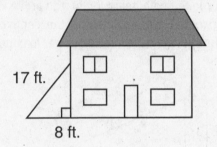

A   5 ft.
B   9 ft.
C   12 ft.
D   15 ft.

You need to find the length of one of the legs. It doesn't matter whether you let a or b represent the missing leg.

$$c^2 = a^2 + b^2$$

$$17^2 = a^2 + 8^2$$

$$289 = a^2 + 64$$

$$225 = a^2$$

$$15 = a$$

The ladder reaches 15 feet up the side of the house. Therefore, answer choice (D) is correct.

Directions: Read the passage below and answer the question that follows.

Drop into Oblivion!

The world record for the steepest drop on a roller coaster is held by the Oblivion in Staffordshire, England. From the top of the first hill, riders fall downward at an incredible 87.5° angle. That's almost straight down! At the bottom of the first 60-meter (197 foot) drop, riders are plummeted deep underground into a 30-meter-deep (98-foot) hole. Oblivion cost $19 million to build. The two-minute ride covers 373 meters (1,223 ft.) of track, attaining a maximum speed of 110 kilometers per hour (68 miles per hour). Riders are subjected to a vertical force of 4.5 Gs.

Look at the diagram below, then solve the sample question on the following page.

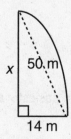

▶ The first drop on a roller coaster is pictured in the figure on the previous page. If the cars cover 50 meters of track and a horizontal distance of 14 meters, how high is the drop (x)?

A   36 m

B   48 m

C   50 m

D   52 m

Know It All Approach

Use the **Know It All Approach** to help you find the height of the drop. Read the question carefully and make note of the words and information you need in order to find the correct answer. All the information you need is contained in the diagram. In the right triangle shown, the hypotenuse is 50 meters and one leg is 14 meters. Your task is to find the length of the other leg.

Next, answer the question. Simply plug the given values into the formula for the Pythagorean theorem.

$$c^2 = a^2 + b^2$$

$$50^2 = a^2 + 14^2$$

$$2{,}500 = a^2 + 196$$

$$2{,}304 = a^2$$

$$48 = a$$

The vertical drop equals 48 meters.

Always remember to check your answer. The easiest way to check your answer is to place it in the formula.

$$c^2 = a^2 + b^2$$

$$50^2 = 48^2 + 14^2$$

$$2{,}500 = 2{,}304 + 196$$

$$2{,}500 = 2{,}500$$

Answer choice (B) is correct.

Directions: Read the passage below and answer the questions that follow.

Dive!

The record for the world's deepest submarine dive is held by the Japanese research submarine *Shinkai 6500*. On August 11, 1989, it achieved a depth of 6,526 meters (21,414 feet) in the Japan Trench off the coast of Sanriku, Japan. The minisub is 9.5 meters long, 2.7 meters wide, and 3.2 meters high (31 feet by 9 feet by 10.5 feet). It weighs 26 tons and has a maximum speed of 2.5 knots. It can maintain life support for its crew of three for 120 hours. In 1999, *Shinkai* made a dive to the Lo`ihi Seamount, an active volcano on the seafloor near Hawaii. The summit of the underwater volcano rises to 969 meters (3,178 feet) below sea level. Seismologists asked the crew to gather seismic data, as Lo'ihi is known for producing frequent earthquakes.

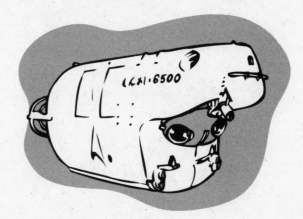

1. A submarine dives at an angle that measures $x°$. Which of the following expressions could be used to represent the measure of its supplement?

   A $180 + x$
   B $90 + x$
   C $180 - x$
   D $90 - x$

2. In the diagram below, what is the measure of ∠QRS?

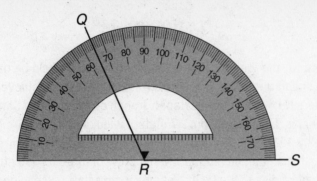

A 65°
B 75°
C 115°
D 125°

3. What is the value of *n* in the figure below?

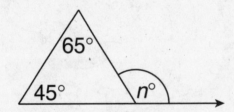

A 45°
B 65°
C 70°
D 110°

Know It All! High School Math

Directions: Read the passage below and answer the questions that follow.

Walk the Line

On May 7, 2002, Chinese acrobat Ahdili broke the record for the longest time spent on a tightrope. He lived on a wire suspended 35 meters (115 feet) above the ground for 22 days. During that time, he wore out three pairs of boots, covering a distance of 200 kilometers (124 miles) on the tightrope. Guinness World Records also certified that Ahdili had set a new record for the longest time (8 hours, 12 minutes) continually walking on a tightrope.

4. In the diagram below, a tightrope is stretched from the roof of one building (160 feet tall) to the roof of another building (110 feet tall). The two buildings are 120 feet apart. What is the length of the tightrope? **Show all work.**

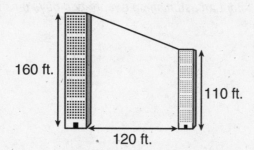

160 ft.

110 ft.

120 ft.

Subject Review

In Chapter 15, you learned how to measure an angle using a protractor. You also learned about many different kinds of angles and their relationship to each other. Lastly, you discovered the Pythagorean theorem for right triangles and how it can be used to find the length of an unknown side.

But all that pales next to the cool facts you've picked up along the way.

What is the angle of descent for the world's steepest roller coaster?
As of 2003, the 87.5° angle of descent at the Oblivion is the steepest coaster in the world.

What is the record for the world's deepest submarine dive?
The Japanese research submarine Shinkai 6500 dove to 6,526 meters (21,414 feet).

What is the world record for the most time spent on a tightrope?
Chinese acrobat Ahdili spent an astonishing twenty-two days on a tightrope!

Directions: Read the passage below and answer the questions that follow.

A Giant Lollipop

Jolly Rancher, a division of Hershey Foods Corporation, produced an enormous lollipop on June 25, 2002. It took a group of more than 50 Hershey employees more than three months to design and build it. It is a scale model of their regular-size Jolly Rancher cherry lollipop. Nearly 16 feet high, including the stick, it is 62.8 inches in diameter and 18.9 inches thick. It weighs 4,106 pounds, nearly 1,000 pounds more than the previous record-holding lollipop.

The giant lollipop is the equivalent of 107,154 regular-sized lollipops. It has been on display at Hershey's Chocolate World Visitors Center in Hershey, Pennsylvania.

This two-ton lollipop contains
6,857,854 calories, but it is fat free!

1. The lollipop shown in the diagram below has a diameter of 6 centimeters and a thickness of 1 centimeter.

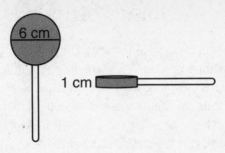

Part A   Find the total surface area of the lollipop. **Show all work**.

_____

Part B   Find the volume of the lollipop. **Show all work**.

_____

2. What is the measure in degrees of each interior angle of the regular polygon shown below?

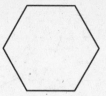

A   60°
B   120°
C   135°
D   720°

3. A man 1.75 meters tall casts a shadow 5.25 meters long. At the same time, a flagpole casts a shadow 120 meters long. How tall is the flagpole?

_____

4. A rectangular room that is 15 feet long has a perimeter of 48 feet. How many square feet of floor tiles are needed to cover the floor?

A   24 ft.²
B   48 ft.²
C   96 ft.²
D   135 ft.²

5. What is the value of *x* in the figure below?

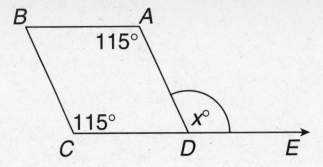

A   65°
B   75°
C   115°
D   125°

Chapter 16: Measurement

When did a car break the sound barrier on land for the first time?

What was unusual about the year 1752?

Who was the youngest person to break the sound barrier?

U.S. Customary Units of Measure

The common units used on a daily basis in the United States are part of the English system of measurement. This system evolved over hundreds of years, during which people used common, everyday objects to measure things. A foot was the length of a person's foot. A cup was the volume of liquid an ordinary cup could hold. A mile was the distance covered by walking a thousand paces. Legend has it that a yard was the distance from King Edward's nose to his extended arm and fingertips. Obviously, this was not a very efficient system. Feet come in different sizes, as do cups and noses. Eventually, all parts of the country agreed on a set of standards. Undoubtedly, you will recognize most of them listed in the table below.

The U.S. Customary System of Measurement

Length
1 foot (ft.) = 12 inches (in.)
1 yard (yd.) = 3 ft.
1 mile (mi.) = 5,280 ft.

Volume
1 cup = 8 fluid ounces (fl. oz.)
1 pint (pt.) = 2 cups
1 quart (qt.) = 2 pt.
1 gallon (gal.) = 4 qt.

Weight
1 pound (lb.) = 16 ounces (oz.)
1 ton = 2,000 lb.

Some test items require you to convert from one unit to another. Examine the sample question on the following page.

▶ Last week, Erica's bean plant was 22 inches tall. It has grown another 7 inches since then. How tall is the bean plant now?

A 1 ft. 10 in.
B 2 ft. 2 in.
C 2 ft. 5 in.
D 2 ft. 9 in.

If the bean plant was 22 inches tall, and it grew another 7 inches, then it is 29 inches tall now. However, that answer does not appear among the answer choices. You need to convert 29 inches to feet and inches. Since 1 foot equals 12 inches, 2 feet equals 24 inches. The bean plant is 2 feet tall plus 5 inches leftover. Answer choice (C) is correct. Try another sample question.

▶ Henry is planning a barbecue with his friends. He figures he will need to make about 60 hamburgers. If it takes 4 ounces of ground beef to make each hamburger, how many pounds of ground beef should he buy?

_____

The first step is to determine how many ounces of ground beef are needed. If you multiply 60 hamburgers times 4 ounces, you get 240 ounces. However, the question asks, "How many pounds of ground beef should he buy?" You need to convert 240 ounces to pounds. Since 16 ounces equal 1 pound, you need to divide: $240 \div 16 = 15$. Henry should buy 15 pounds of ground beef.

Henry should buy 15 pounds of ground beef.

The Metric System

Most of the world uses the metric system of measurement. Scientists in Europe created this system about 200 years ago. The idea was to make measurement easier. Instead of multiplying by 12, or 3, or some other arbitrary number, all units in the metric system are related to each other by powers of 10. Look at the table on the following page and see if you are familiar with any of the metric units.

The Metric System of Measurement

Length
1 centimeter (cm) = 10 millimeter (mm)
1 meter (m) = 100 cm
1 kilometer (km) = 1,000 m

Weight
1 centigram (cg) = 10 milligram (mg)
1 gram (g) = 100 cg or 1,000 mg
1 kilogram (kg) = 1,000 g

Volume
1 centiliter (cl) = 10 milliliter (ml)
1 liter (L) = 100 cl or 1,000 ml
1 kiloliter (kl) = 1,000 L

To convert from one unit to another, all you need to do is multiply or divide by a power of 10. To get an idea of how this works, try the sample question below.

▶ Lorna took her dog to the vet. Last year, the dog weighed 44 kilograms. This year, it weighs 3,450 grams more than that. What is the dog's weight in kilograms this year?

A   3.45 kg
B   40.55 kg
C   44.345 kg
D   47.45 kg

You cannot work with both kilograms and grams. You need to make some kind of conversion so that you are working with only one unit—either grams or kilograms. To convert from kilograms to grams, **multiply** by 1,000, or move the decimal point three places to the **right**. To convert from grams to kilograms, **divide** by 1,000, or move the decimal point three places to the **left**. Here's one way to solve the above problem.

44 kg = 44,000 g

44,000 + 3,450

$\frac{47,450}{1,000}$

47.45 kg

Answer choice (D) is correct.

Try another sample question.

▶ Which of the following represents the greatest distance?

A   0.6 km
B   750 m
C   8,000 cm
D   50,000 mm

You need to compare the four answer choices to see which is the greatest. In order to do that, you need to convert them to one unit. Since it is midway in size, the best units to use is the meter.

A   0.6 km = 600 m
B   750 m = 750 m
C   8,000 cm = 80 m
D   50,000 mm = 50 m

Now that you have converted all the answer choices to meters, it is obvious that the greatest distance is 750 m. Answer choice (B) is correct.

Comparing Customary and Metric Units

Normally, test items do not ask you to convert between systems. Nevertheless, you should have some idea of how big a meter, a kilogram, or a liter is. Study the table below, which gives approximate equivalents.

Customary to Metric	Metric to Customary
Length 1 m = 39.37 in. 1 km = 0.62 mi.	**Length** 1 in. = 2.54 cm 1 mi. = 1.609 km
Weight 1 kg = 2.2 lb.	**Weight** 1 lb. = 0.453 kg
Volume 1 L = 1.057 qt.	**Volume** 1 qt. = 0.946 L

Notice that a meter (39.37 inches) is slightly longer than a yard (36 inches). Most rulers come with inches on one side and centimeters on the other. How long is a 6-inch ruler in centimeters? Usually, a 6-inch ruler has about 15 centimeters (2.54 × 6) on the metric side. If you weigh 50 kilograms, what is your weight in pounds? A person weighing 50 kilograms would weigh about 110 pounds (50 × 2.2). Notice too that a liter is just slightly larger than a quart.

To see how you might be asked to use this information, check out the sample question below.

▶ A car is 6 meters long. How long is the car in feet and inches?

A   18 ft.
B   18 ft. 8 in.
C   19 ft.
D   19 ft. 8 in.

You need to make a conversion between meters and feet. A meter is 39.37 inches.

$$6 \text{ m} = 6 \cdot 39.37 \text{ in.}$$

$$6 \text{ m} = 236.22 \text{ in.}$$

$$6 \text{ m} = 19 \text{ ft. } 8 \text{ in.}$$

Therefore, 6 meters is 19 feet 6 inches, and answer choice (D) is correct.

Estimating

If you are familiar with the length, weight, or volume of common objects in both the customary system and the metric system, you can estimate the size of unknown objects. For example, can you estimate the height of your bedroom door? Most likely, the door is about 2 meters, or 6 feet 6 inches high. Knowing that, try to solve the sample question on the next page.

► Which of the following could be the area of a school classroom?

A   9 m²

B   16 ft.²

C   450 ft.²

D   60,000 cm²

You could probably convert all of these answer choices to one unit, like square feet, but why bother? Use your ability to estimate.

For the area to be 9 square meters, the classroom could measure 3 meters by 3 meters. That's about 10 feet by 10 feet. Even a small bedroom would be bigger than that, so you can eliminate choice (A). For the area to be 16 square feet, the classroom could measure 4 feet by 4 feet. That is too small, so you can rule out choice (B). For the area to be 450 square feet, the classroom could measure a little more than 20 feet by 20 feet. That seems like a reasonable size for a classroom, so keep choice (C) in mind. For the area to be 60,000 square centimeters, the classroom could measure 300 centimeters by 200 centimeters. That's the same as 3 meters by 2 meters, which is even smaller than choice (A). Hence, you can get rid of choice (D). The only reasonable answer, choice (C), is correct. Try another sample question.

► About how much water would it take to fill a small child's inflatable swimming pool?

A   20 L

B   50 gal.

C   128 fl. oz.

D   3,375 cm³

Again, to convert among liters, gallons, fluid ounces, and cubic centimeters would be tedious and time-consuming. Try to arrive at the correct answer by estimation.

A volume of 20 liters would be almost the same as 20 quarts, which is 5 gallons. That would be a lot to drink, but it would barely make a puddle inside a child's pool. Hence, you can eliminate choice (A). A volume of 50 gallons sounds more reasonable, so hold a place for choice (B). A volume of 128 fluid ounces is equal to 1 gallon. That would barely wet the pool, so you can get rid of choice (C). A volume of 3,375 centimeters would be equal to a cube measuring 15 centimeters on each side. That volume of water would not even come close to filling the child's pool. You can throw out choice (D) As you suspected, answer choice (B) is correct.

Directions: Read the passage below and answer the question that follows.

Super Car

In October 1997, in the Black Rock Desert in Nevada, the British *Thrust SSC* became the first vehicle to break the sound barrier on land. On October 13, Royal Air Force pilot Andy Green made two runs across the desert at 760.135 miles per hour and 763.168 miles per hour—both supersonic! However, the British team was disqualified from an official record because it took one minute more than the required sixty minutes to complete both runs.

Undaunted, the team successfully broke the official world land speed record two days later. On October 15, the jet-powered car achieved a speed of 759.333 miles per hour on its first run. On its second run, thirty minutes later, the first official supersonic land speed record was set when it zoomed by at 766.109 miles per hour.

Just as it had done two days earlier, the *Thrust SSC* generated a thunderous sonic boom during each run. The speed of sound varies according to temperature, humidity, and altitude. On the morning of the 15th, officials calculated that it was 748.111 miles per hour.

▶ Which of the following would be the same speed as 720 miles per hour?

    A   0.2 miles per second

    B   1 mile per second

    C   30 miles per minute

    D   72 miles per minute

Know It All Approach

First, read the question carefully and make note of the words and information that you need in order to find the correct answer. You need to compare miles per hour, miles per minute, and miles per second. You are trying to figure out what speed in miles per second or miles per minute is equivalent to 720 miles per hour.

Now, answer the question. Because a minute is less than an hour and greater than a second, it would be a good idea to convert all the speeds to miles per minute.

A	0.2 miles per second ($\times$ by 60)	12 miles per minute
B	1 miles per second ($\times$ by 60)	60 miles per minute
C	30 miles per minute	30 miles per minute
D	72 miles per minute	72 miles per minute

$$\frac{720 \text{ miles per hour}}{60} = 12 \text{ miles per minute}$$

Double-check your answer. Wherever you multiplied, divide to check your answer. Wherever you divided, multiply to check your answer.

    A $12 \div 60 = 0.2$

    B $60 \div 60 = 1$

            $12 \times 60 = 720$

Read all of the answer choices, and eliminate those you *know* are incorrect. The only answer choice that agrees mathematically with 720 miles per hour is answer choice (A).

Directions: Read the passage below and answer the questions that follow.

Missing Days?

In 1752, an act of Parliament decreed that the Gregorian calendar should replace the Julian calendar in both England and the American colonies. Due to faulty calculations by Roman astronomers in 46 B.C., the difference between the two calendars had grown to 11 days. Accordingly, it was ruled that the day after Wednesday, September 2, would become Thursday, September 14! When the change was enacted, people became puzzled. Local gossip suggested that workers were losing 11 days of pay, and everyone was losing 11 days of their lives. They rioted in the streets of London in protest. Eventually, people accepted the new calendar, which we still use to this day. The year 1752 remains the only year that was 354 days long.

September 1752						
Sun	Mon	Tue	Wed	Thu	Fri	Sat
		1	2	14	15	16
17	18	19	20	21	22	23
24	25	26	27	28	29	30

1. In 1752, September had only 19 days. How many minutes did it have?

   A   456

   B   1,140

   C   1,440

   D   27,360

2. Which of the following could be the weight of a small dog?

   A   20 kg

   B   60 lb.

   C   320 oz.

   D   1,200 g

3. When you set up your stereo system, you discover that you need a 9-foot electrical extension cord. When you get to the store, all the extension cords are labeled in meters. Which of the following lengths should you buy?

A   1 m
B   3 m
C   6 m
D   10 m

4. A baby weighed 104 ounces at birth. Two weeks later, the baby had gained 20 ounces in weight. How much did the baby weigh at two weeks old?

A   3 lbs. 28 oz.
B   5 lbs. 4 oz.
C   6 lbs. 4 oz.
D   7 lbs. 12 oz.

Directions: Read the passage below and answer the questions that follow.

Fast Learner

On July 12, 1994, eleven-year-old Katrina Mumaw became the youngest person to break the sound barrier. Katrina got hooked on flying at age three and has been dogfighting in military-style airplanes since she was eight. Before she could break the sound barrier, Katrina first had to prove her flying ability to her instructor, Russian pilot Vladimir Danilenko, by flying a jet trainer. Once he was assured of her competence, Danilenko and his young pupil boarded a Russian MiG-29UB fighter.

Following takeoff, Katrina and Vladimir flew over the test range where Katrina piloted the jet to a speed of Mach 1.3 (approximately 940 miles per hour). In the future, Katrina hopes to win appointment to the U.S. Air Force Academy and eventually became an astronaut.

5. Which of the following would be the same speed as 180 kilometers per hour?

    A   3 kilometers per minute
    B   3 kilometers per second
    C   30 kilometers per minute
    D   108 kilometers per second

6. Which of the following containers will hold the most liquid?

    A $\frac{1}{2}$ gal. jug
    B   2 L bottle
    C   16 oz. cup
    D   1,000 ml flask

7. A ruler is 12 inches long. About how long is the ruler in centimeters?

    A   5 cm
    B   15 cm
    C   30 cm
    D   36 cm

Subject Review

In Chapter 16, you learned about two systems of measurement: the English system and the metric system. You were shown how to convert from one unit to another within each system. You gained some sense of how measurements in one system are related to those in the other system. You also learned how estimate the measurements of objects.

But that's not all! You now know the following:

When did a car break the sound barrier on land for the first time?
In October 1997, the British Thrust SSC became the first vehicle to break the sound barrier on land.

What was unusual about the year 1752?
It was only 354 days long.

Who was the youngest person to break the sound barrier?
Eleven-year-old Katrina Mumaw on July 12, 1994, flew a Russian jet 940 mph!

Chapter 17: Mean, Median, and Mode

What is the world record for the most custard pies thrown in three minutes?

How big was the world's largest iceberg?

How much did the world's most valuable painting sell for at auction?

Mean, Median, and Mode

Statistics is the branch of mathematics that deals with collecting, organizing, and analyzing data. **Data** are facts or pieces of information. Examine the set of data listed below.

11, 20, 13, 7, 21, 5, 12, 17, 11

The data represent the number of points an eleventh-grade girl scored in each of her first nine basketball games this season. What do you do now? You already have the data, so there is no collecting to be done. You could organize the data into a table or graph. Since you will learn about graphs in the next chapter, make a table of the data for now.

Game	Points
1	11
2	20
3	13
4	7
5	21
6	5
7	12
8	17
9	11

The data have been collected and organized. The next step is to analyze the data. There are a number of ways to describe this or any other set of data. One way is to find the mean. The **mean** is the sum of the numbers in a set of data divided by the number of pieces of data. The mean is what people usually think of when they say "average." Find the mean of the set of data listed in the table.

$$\text{mean} = (11 + 20 + 13 + 7 + 21 + 5 + 12 + 17 + 11) \div 9$$

$$\text{mean} = 117 \div 9$$

$$\text{mean} = 13$$

To find the mean, you add up all the scores and then divide by the number of scores. The mean for this set of data is 13. This means that the girl averages 13 points per game.

Another way to describe a set of data is to find the median. The **median** is the middle number in a set of data when the data are arranged in numerical order. Find the median of the set of data listed in the table.

$$5, 7, 11, 11, \mathbf{12,} 13, 17, 20, 21$$

To find the median, you need to list the scores from least to greatest. The median is the middle number. In this set of data, the median is 12 because there are four scores lower than 12 and four scores higher than 12.

If you had an even number of scores in the data set, the median would be the mean of the two middle numbers. For example, in the data set 2, 3, **6, 10,** 17, 20, the median is the mean of the two middle numbers, that is, $(6 + 10) \div 2$, or 8.

A third way to describe a set of data is to find the mode. The **mode** is the number or item that appears most often in a set of data. Find the mode of the set of data listed in the table.

$$\mathbf{11,} 20, 13, 7, 21, 5, 12, 17, \mathbf{11}$$

In most cases, the mode is easy to find. As you can see, the number 11 appears most often in this set of data. Therefore, the mode is 11. Some data sets do not have a mode because no two numbers in the set are identical. Other data sets have two or more modes. To check your understanding of these ways to describe a set of data, try the sample question below.

▶   Nelson bowled three games after school. His scores were 143, 127, and 165. What was his mean score?

   A   127

   B   143

   C   145

   D   165

Know It All! High School Math

To find the mean, you need to add up all the scores and then divide by the number of scores.

mean $= (143 + 127 + 165) \div 3$

mean $= 435 \div 3$

mean $= 145$

Nelson's mean score was 145. Therefore, answer choice (C) is correct. Try another sample question.

▶ For a biology project, Nina recorded the time it took her pet gerbil to run through a maze. She did this every day for a week. The results are shown in the table below.

Day	Time (sec.)
Mon	26.3
Tue	29.6
Wed	22.9
Thu	18.2
Fri	19.8
Sat	11.4
Sun	11.8

What is the median time, in seconds, for the gerbil running through the maze?
    A   18.2
    B   19.8
    C   20
    D   22.9

To find the median, you need to list the scores from least to greatest. Then, find the middle number.

11.4, 11.8, 18.2, **19.8**, 22.9, 26.3, 29.6

The middle number, 19.8, is the median. Therefore, answer choice (B) is correct. Here's one more sample question.

▶ What is the mode of the data set shown below?

4, 7, 8, 4, 11, 7, 9, 4, 5

A 4
B 5
C 7
D 8

To determine the mode, you need to find the number that appears most often in the data set. The number 4 appears three times. Therefore, the mode is 4, and answer choice (A) is correct.

So far, this statistics stuff is pretty easy. However, math teachers love to complicate matters. Consider the next sample question.

▶ Ramie's test scores in social studies this marking period are 88, 95, 91, 87, and 84. What grade would he need to get on his next test in order to make his average (mean) equal 90?

A 75
B 89
C 91
D 95

Notice that the question does not ask you to find the mean, median, or mode. It asks you what grade on the next (sixth) test would make the average (mean) equal 90. For six test grades to have a mean of 90, the total of the six grades must be 6×90, or 540. The first step is to find the sum of the grades so far, that is, the sum of the first five test grades.

$$88 + 95 + 91 + 87 + 84 = 445$$

Next, subtract the sum from the needed total of 540.

$$540 - 445 = 95$$

Ramie would need to get a grade of 95 on his next test to bring his average up to 90. Therefore, answer choice (D) is correct. You can double-check your answer by finding the mean of all six test grades.

$$\text{mean} = (88 + 95 + 91 + 87 + 84 + 95) \div 6$$

$$\text{mean} = 540 \div 6$$

$$\text{mean} = 90$$

Directions: Read the passage below and answer the question that follows.

Pie in the Face

In April 2000, a new record was set for the world's biggest custard pie fight. At London's Millennium Dome, twenty participants threw 3,312 pies in three minutes, breaking the previous record of 3,076 set in 1998. Each participant was assigned a partner who fed the thrower a continuous supply of pies.

To make the gooey ammunition, kitchen workers mixed more than a thousand pounds of custard powder with one thousand liters of water in a half dozen cement mixers. The yellow ooze was then poured into six thousand pie pans.

▶ Six people participated in a pie-throwing contest. The table below lists the number of pies each threw in three minutes. What is the mean number of pies thrown?

Participant	Pies Thrown
1	146
2	152
3	169
4	154
5	172
6	149

A   140
B   153
C   157
D   188

Know It All Approach

Use the **Know It All Approach** to conquer this question. First, read the question carefully and make note of the words and information you need in order to find the correct answer. All the information you need is contained in the table. Your task is to find the mean of the data set. Once you've got the data you need, answer the question. To find the mean, you need to add up all the numbers and then divide by the number of numbers.

mean = (146 + 152 + 169 + 154 + 172 + 149) ÷ 6

mean = 942 ÷ 6

mean = 157

Now that you've found your answer, check your work. Because you divided to get your answer, you should multiply to double-check it. Because 6 × 157 = 942, just check your addition to make sure the sum of the numbers is also 942.

Finally, read all the answer choices, and eliminate those you *know* are incorrect. Choice (A) is smaller than any number in the data set, so it could not be the mean. Hence, you can eliminate it. Similarly, choice (D) is larger than any number in the data set, so it could not be the mean either. You can eliminate it as well. Choices (B) and (C) are both reasonable. However, after checking the math, answer choice (C) is correct.

Directions: Read the passage below and answer the questions that follow.

Chill Out

Perhaps the world's largest iceberg was spotted by a NOAA (National Oceanic and Atmospheric Administration) weather satellite in March of 2000. A photograph transmitted from a satellite whose orbit takes it over both poles indicates that the giant iceberg broke free from the Antarctic's Ross Ice Shelf.

Measuring 295 kilometers (183 miles) long and 37 kilometers (23 miles) wide, with a surface area of 11,000 square kilometers (4,250 square miles), it may be the largest iceberg ever recorded. It's as large as a small country.

Oceanographers believe the iceberg will follow one of two paths. Because of its immense size, it may remain intact for a long time, even if it floats to waters that are more temperate. However, most scientists believe it will become trapped by Antarctic currents and remain close to the cold continent.

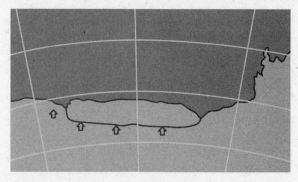

The Antarctic iceberg is the size of a small country.

1. The table below shows the length in miles of five recently spotted icebergs. What is the mean length of these icebergs?

Iceberg	Length (miles)
A	47
B	22
C	183
D	31
E	42

A   31 miles
B   42 miles
C   54 miles
D   65 miles

2. The following data set lists the expected gasoline mileage of nine different automobiles in miles per gallon. What is the median of this set of data?

38.4, 17.2, 23.6, 30.4, 36.1, 19.5, 32.2, 17.0, 28.8

A   23.6 mpg
B   28.8 mpg
C   29.6 mpg
D   30.4 mpg

3. The table below shows the number of students in each age bracket who attend John Adams High School. What is the mode of their ages?

Age	Frequency
13	114
14	189
15	218
16	235
17	221
18	88

A $15\frac{1}{2}$

B 16

C $16\frac{1}{2}$

D 17

Directions: Read the passage below and answer the questions on the following pages.

Pricey Paintings

Picasso is ranked as the artist with the most works sold for more than a million dollars. As of 2003, he has sold 298 works, the value of sales totaling $1,399,203,108. The mean value of a Picasso work is a little under $4,700,000. The most highly insured work of art is the *Mona Lisa* by Leonardo da Vinci. Before being moved to Washington, DC, and New York, NY, for exhibition in late 1962, it was assessed by independent insurance companies at $100 million.

However, the record for the most valuable painting ever sold belongs to *Portrait of Doctor Gachet* by Vincent van Gogh. It sold at Christie's auction house in New York City for $82.5 million on May 15, 1990. The date was just ten weeks short of the one hundrendth anniversary of his death.

4. The prices of the ten most expensive paintings are listed in the table below. What is the median price of these paintings?

10 Most Expensive Paintings, as of 2003

Rank	Price (millions)	Artist
1	$82.5	Van Gogh
2	$78.1	Renoir
3	$71.5	Van Gogh
4	$65	Picasso
5	$60	Cézanne
6	$51.65	Picasso
7	$49.5	Picasso
8	$49	Van Gogh
9	$48	Picasso
10	$47.85	Picasso

A $51.6 million
B $55.825 million
C $60 million
D $60.3 million

5. Mr. Archer, a car salesman, wants to average thirty cars sold per month for six months. The table shows how many cars he has sold over the last five months.

Month	Cars Sold
Jan.	15
Feb.	22
March	35
April	39
May	29
June	?

How many cars does Mr. Archer need to sell in June in order to achieve his goal?

A   28
B   29
C   35
D   40

6. What is the mode of the data set shown below?

77, 90, 88, 88, 74, 82, 88, 72, 88, 75, 69

A   81
B   82
C   88
D   89

Subject Review

In Chapter 17, you learned how to analyze data by calculating the mean, the median, and the mode. The mean is the sum of the numbers in a set of data divided by the number of pieces of data. The median is the middle number in a set of data when the data are arranged in numerical order. The mode is the number or item that appears most often in a set of data.

You also know some strange facts.

What is the world record for the most custard pies thrown in three minutes?

3,312 pies were tossed in a giant food fight. It probably took more than three minutes to clean up!

How big was the world's largest iceberg?

A chunk of Antarctica measuring 295 kilometers (183 miles) long and 37 kilometers (23 miles) wide broke off in March 2000.

How much did the world's most valuable painting sell for at auction?

A Japanese businessman paid $82.5 million for a single painting in 1990.

Chapter 18: Graphs and Data

What is the world skateboard speed record?

Which parasite is the most beneficial to humans?

Was the sinking of the ocean liner "Titanic" foretold?

Graphs and Data

In the last chapter, you learned how to analyze data. In this chapter, you will learn various ways of displaying data. You are already familiar with data tables, which allow you to display data in a neat and organized manner. However, they lack the visual impact of a graph. Examine the diagrams below.

Country	Estimated Population (millions)
Algeria	35
Angola	13
Cameroon	14.5
Egypt	65
Ethiopia	58.5
Ghana	22
Kenya	38.5
Morocco	36.5
Nigeria	160
South Africa	47
Sudan	33
DROC	52.2

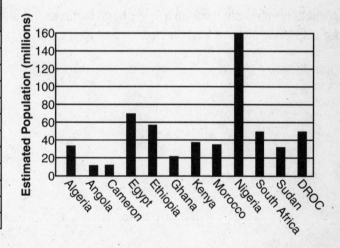

Both the table and the graph display the same information. However, the overpopulation of Nigeria is displayed more dramatically on the graph because of the extreme height of the bar. A **bar graph** uses bars to compare statistics. It is useful for showing information collected by counting.

Next, look at another type of graph.

Solid Waste	Percent
Yard Waste	18%
Food	8%
Plastic	7%
Paper	37%
Metal	10%
Glass	10%
Other	10%

Again, both the table and the graph display the same information. However, the graph displays the data in a visual way, while it also shows the numbers. A **circle graph** divides a circle into pieces whose relative sizes correspond to the data.

A circle graph is often called a "pie chart" because the pieces look like slices of a pie. It is useful for displaying parts of a whole. Notice that all the percents (the parts) add up to 100% (the whole).

You are probably familiar with the next type of graph.

U.S. Whole Milk Consumption

Year	Gallon per capita
1965	30
1970	25
1975	20
1980	16
1985	14
1990	11
1995	8

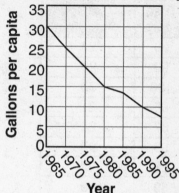

U.S. Whole Milk Consumption

The data in the table show how the consumption of whole milk by U.S. residents has declined over the last thirty years. The numbers in the table tell you it has declined. However, the decline is much more obvious on the graph. You do not need to look at the numbers; just look at the downward trend of the line on the graph.

A **line graph** uses points connected by a line to show the relationship between two variables. The independent variable goes on the horizontal or *x*-axis, and the dependent variable goes on the vertical or *y*-axis. A line graph is especially suited for showing changes over time.

The next and last graph you need to consider is a close relative of the line graph. The data below is based on sixteen students in one class.

Test Score	Hours Spent Studying
98	27
96	24
100	30
95	21
87	28
95	20
88	22
89	19
78	16
84	18
81	14
72	10
65	11
68	8
67	5
52	3

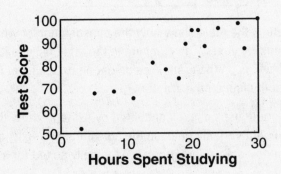

A student group was wondering whether there was any connection between the hours spent studying and test scores.

A **scatterplot** shows the relationship between two sets of data. Like a line graph, it compares the association between two variables. However, unlike a line graph, you do not connect the points. What you do is look for a trend.

Look at the three scatterplot diagrams below.

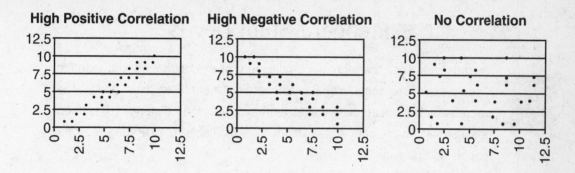

If the points form a pattern that moves up and to the right, there is a **high positive correlation** between the variables.

If the pattern runs down and to the right, there is a **high negative correlation** between them.

If there is no apparent pattern, the variables are unrelated and their is **no correlation**.

Refer to the scatterplot on the previous page. You should notice that there appears to be a positive correlation between hours spent studying and test scores. That is, as the time spent studying increases, the test scores also increase. It does not *necessarily* mean that more hours studying cause higher test scores. However, it does suggest the general trend that more studying results in better test scores.

Skateboard Stunt Devils

No one can deny the increasing popularity of skateboarding and extreme sports. Here are some world skateboarding records to think about. The records for longest air distance (65 feet) and highest air distance (18 feet) on a skateboard were both set by Danny Way in 2002 at the OP King of Skate contest in California.

On September 26, 1998, Gary Hardwick set a speed record of 62.55 miles per hour (100 kilometers per hour) for human-powered skateboarding in a standing position. Gary wore an aerodynamic suit and helmet to accomplish the feat. The record speed (70 miles per hour or 112 kilometers per hour) for a rocket-powered skateboard is held by Billy Copeland. Billy mounted eight jetpacks of fuel to the back of his skateboard. The record for the most midair rotations (900° or $2\frac{1}{2}$ full rotations) on a skateboard was set by legendary Tony Hawk at the 1999 X Games.

▶ Study the chart below, and then use the data to answer the question that follows.

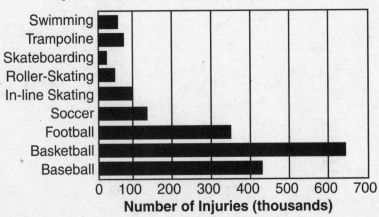

Sports-Related Injuries in the U.S. in 1996

Based on the data in the graph, about how many more injuries occurred playing baseball than occurred skateboarding?

A   400

B   600

C   400,000

D   600,000

Know It All Approach

Using the **Know It All Approach** will make answering this question painless. Start by reading the question carefully. Make note of the words and information you need in order to find the correct answer. The information you need is contained in the graph. Your task is to find the difference between the number of baseball injuries and the number of skateboarding injuries.

Examine the graph. Once you've figured out the question and the graph, answer the question. You should realize that each space on the graph represents 100,000 injuries. Since the bar for baseball is a little less than halfway between the 400,000 and 500,000 lines, you can estimate that the number of baseball injuries is about 440,000. Similarly, the bar for skateboarding is a little less than halfway between the zero line and the 100,000 line. Hence, you can estimate that the number of skateboarding injuries is about 35,000 or 40,000. By subtracting, you can figure out that the difference is about 400,000.

Don't forget to double-check your answer. There is no simple method you can use in this case. Just make sure that you looked at the appropriate bars on the graph and that you read them accurately.

Increase your chances of finding the correct answer by reading all the answer choices and eliminating those that you know are incorrect. You should realize that choices (A) and (B) are designed to trick you in case you failed to notice that the numbers on the graph are in thousands. Hence, you can eliminate both. Similarly, choice (D) is designed to trick those who mistakenly read the bar for basketball instead of baseball. Thus, you can get rid of choice (D) as well. The only answer choice remaining, choice (C), is the correct answer.

Directions: Read the passage below and answer the questions that follow.

A Cure That Sucks

During the 1800s, bloodletting was a common medical treatment. Doctors applied leeches to the skin of their sick patients. At the time, many people believed that illnesses were caused by an imbalance of the body's fluids.

Today, leeches still have a role in medicine because of their healing powers. They are primarily used to relieve blood congestion in delicate operations. By draining the blood from bruised areas, they reduce swelling. Leeches also help to restore blood flow to injured body tissues or severed limbs that have been reattached through microsurgery.

Leeches have a suction cup on each end of their bodies. The mouth of a leech is a part of the front suction. Their mouth has three jaws containing more than three hundred teeth! Leech saliva contains a painkiller, an anticoagulant, and an antibiotic. Guinness World Records has certified the leech as the parasite most beneficial to humans.

1. Study the graph below and answer the question on the following page.

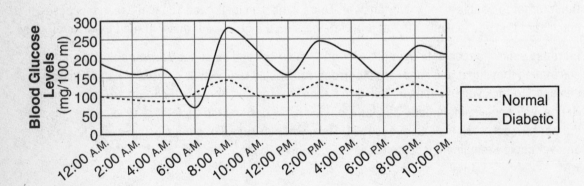

Medical researchers wanted to study how the behavior of leeches is affected by variations in blood glucose levels. The graph below shows typical levels for a normal patient versus a diabetic patient. At 8:00 A.M., approximately how much higher is the blood glucose level of the diabetic patient than that of the normal patient?

A   135 mg/100 ml
B   150 mg/100 ml
C   220 mg/100 ml
D   275 mg/100 ml

2. The circle graph below shows how the 600 students at Jefferson High School travel to school each day. Study the graph, and answer the following question.

Travel to School

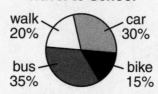

According to the graph, how many students travel by bus?

A   35
B   120
C   180
D   210

3. Draw a scatterplot of the data provided in the table below. Is there a positive correlation, a negative correlation, or no correlation between the variables?

Age	Height (cm)
5	80
7	90
8	110
10	110
10	130
12	140
12	130
13	150
14	170
16	160

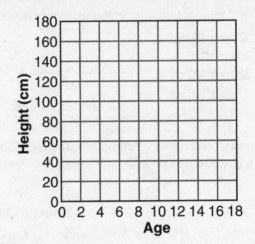

4. Felicia has her own Web site. She made a graph showing the number of people from various foreign countries who visited her Web site.

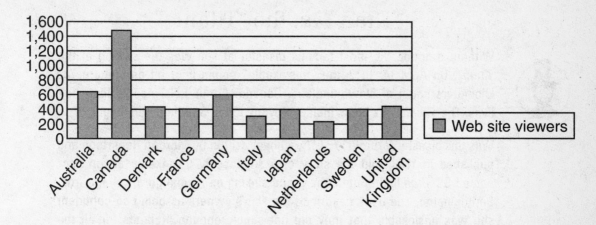

From how many countries did Felicia have 500 or more people visit her Web site?

A   3 countries

B   5 countries

C   6 countries

D   7 countries

Directions: Read the passage below and answer the questions that follow.

SS "Titan" vs. the "Titanic"

Without a doubt, the most famous disaster at sea was the sinking of the *Titanic*. On April 14, 1912, the "unsinkable" ocean liner hit an iceberg 700 kilometers south of Newfoundland, Canada. About 1,500 people lost their lives. The *Titanic* was on its maiden voyage out of Southampton, England.

Was this disaster foretold? *Futility*, a novel written by Morgan Robertson and published in 1898, told the story of a supposedly unsinkable ocean liner named SS *Titan* hitting an iceberg and sinking on its inaugural voyage from Southampton. The book described the ship's owners as being so confident she was unsinkable that they did not supply enough lifeboats for all the passengers. These similarities, and many others as shown in the table below, are really quite amazing.

SS *Titan*		*Titanic*
British	**Country**	British
800 feet	**Length**	882 feet
70,000 tons	**Displacement**	60,250 tons
24 knots	**Top Speed**	24 knots
3	**Propeller**	3
19	**Watertight Bulkheads**	15
3,000	**Capacity**	3,000
2,000	**Actual Passengers**	2,200
24	**Lifeboats**	20
starboard bow	**Damage**	starboard bow
April	**Month**	April

5. The film *Titanic* was one of the top-grossing films of all time, breaking records for both theater attendance and rentals. The graph below compares the amount spent on VHS rentals to the amount spent on DVD rentals for six consecutive years. During which time interval did the amount spent on DVD rentals, compared to VHS rentals, increase the most?

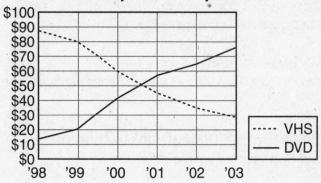

U.S. Rentals per $100 Spent

A   1998 to 1999
B   1999 to 2000
C   2000 to 2001
D   2001 to 2002

Subject Review

In Chapter 18, you learned about a variety of ways to display data. You discovered how to construct and interpret bar graphs, circle graphs, line graphs, and scatterplots. You also learned which type of display is best suited for certain types of data.

And now, the answers to the questions from page 219:

What is the world skateboard speed record?

Gary Hardwick raced 62.55 miles per hour (100 kilometers per hour) on a human-powered skateboard.

Which parasite is the most beneficial to humans?

The leech sucks our blood, but it is also useful in surgery.

Was the sinking of the ocean liner "Titanic" foretold?

Fourteen years before the disaster, a book written by Morgan Robertson told the story of an unsinkable ocean liner named SS Titan hitting an iceberg and sinking on its maiden voyage.

Chapter 19: Probability

What are the odds of winning the lotto?

Which bird can count up to six, recognize seven different colors, and distinguish among five different shapes?

In the future, how might some forms of cancer be detected?

Simple Events

Every time you flip a coin, you perform an action that involves chance. An **outcome** is one possible result of that action. With a coin flip, there are only two possible outcomes: heads or tails. An **event** is a specific outcome. Flipping a coin and having it land on heads is an event. **Probability** is the measure of how likely an event is. It is the chance that a certain event will happen. In math, probability is the ratio of the number of ways a certain event can occur to the number of possible outcomes.

$$P = \frac{\text{number of ways an event can occur}}{\text{number of possible outcomes}}$$

What is the probability of flipping a coin and having it land on heads? The number of ways this event can occur is only one: heads. The number of possible outcomes is two: heads and tails. Therefore, the probability of getting heads is 1 out of 2, or $\frac{1}{2}$. Probability is written as a number between 0 and 1. When it is impossible for an event to occur, the probability is 0. For example, the probability that $100 bills will suddenly fall from the sky is 0. When it is certain that an event will occur, the probability is 1. For example, the probability that tomorrow the sun will rise in the east is 1. Probability can be expressed as a fraction, a decimal, or a percent. The probability of flipping a coin and having it land on heads is $\frac{1}{2}$, or 0.5, or 50%. Try the sample question below.

▶ A board game uses a cube with sides numbered 1 through 6 to determine how many spaces each player moves. What is the probability of rolling a 5?

A $\frac{1}{6}$

B $\frac{1}{5}$

C $\frac{1}{2}$

D $\frac{5}{6}$

There is only one way this event (rolling a 5) can occur. However, there are six possible outcomes. Therefore, $P = \frac{1}{6}$, and answer choice (A) is correct.

Compound Events

A **simple event** involves one action, such as flipping one coin, drawing one card, or rolling one number cube. A **compound event** involves two or more actions. For example, flipping **two** coins would be a compound event. What is the probability that both coins will land on heads? You already know that the probability of one coin landing on heads is $\frac{1}{2}$. How do you find the probability of two coins landing on heads? The probability of a compound event is the **product** of the probabilities of each simple event.

$$P(A \text{ and } B) = P(A) \cdot P(B)$$

Therefore, the probability of two coins both landing on heads is $\frac{1}{2} \cdot \frac{1}{2}$, or $\frac{1}{4}$. Here's another sample question for you.

▶ Shown in the figure below are two fair spinners used in a game. What is the probability of both spinners landing on red?

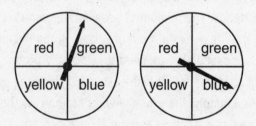

A $\frac{1}{2}$

B $\frac{1}{4}$

C $\frac{1}{8}$

D $\frac{1}{16}$

Know It All! High School Math

Because each spin is a simple event, two spins make a compound event. The probability of landing on red for each spin is $\frac{1}{4}$. Hence, the probability that both spins will land on red is $\frac{1}{4} \cdot \frac{1}{4}$, or $\frac{1}{16}$. Therefore, answer choice (D) is correct.

Flipping two coins or spinning two spinners are **independent events**, that is, the outcome of one event does not affect the outcome of the other event. Some events are **dependent events**, that is, the outcome of one event **does** affect the outcome of the other event. Consider a standard deck of fifty-two cards. Suppose you pick a card, do **not** put it back in the deck, and then pick another card. What is the probability that you will pick two aces? On the first pick, the probability of picking an ace is $\frac{4}{52}$, or $\frac{1}{13}$. However, for the second pick, there are only fifty-one cards remaining in the deck, three of which are aces. The probability of picking an ace on the second pick is $\frac{3}{51}$, or $\frac{1}{17}$. Therefore, the probability of picking two aces is $\frac{1}{13} \cdot \frac{1}{17}$, or $\frac{1}{221}$. See if you can figure out the answer to the sample question below.

▶   A candy jar contains 5 red gumballs, 4 green gumballs, and 3 yellow gumballs. Martin takes out 1 gumball at random and does **not** replace it in the jar. He then takes out another gumball at random. What is the probability that both gumballs chosen are red?

A $\frac{1}{3}$

B $\frac{5}{12}$

C $\frac{5}{33}$

D $\frac{25}{144}$

The probability of the first gumball chosen being red is $\frac{5}{12}$. Because the red gumball is not replaced, the probability of the second gumball chosen being red is $\frac{4}{11}$. Therefore, the probability that both gumballs chosen are red is $\frac{5}{12} \cdot \frac{4}{11}$, or $\frac{20}{132}$, which can be reduced to $\frac{5}{33}$. Therefore, answer choice (C) is correct.

Occasionally, compound events can get very complicated. Pretend a young boy is deciding what to wear. He looks in his closet and sees two shirts (one blue and one black), three pairs of pants (one gray, one tan, and one denim), and two pairs of shoes (one pair of sneakers and one pair of boots). How many different outfits can he put together? What the question is really asking is how many possible outcomes are there? A **tree diagram** is a diagram used to show the total number of possible outcomes.

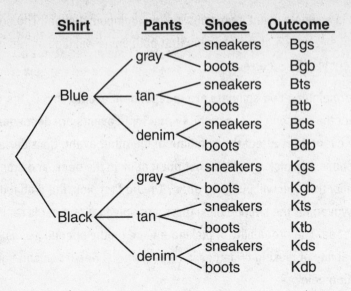

Shirt	Pants	Shoes	Outcome

Blue
- gray
  - sneakers — Bgs
  - boots — Bgb
- tan
  - sneakers — Bts
  - boots — Btb
- denim
  - sneakers — Bds
  - boots — Bdb

Black
- gray
  - sneakers — Kgs
  - boots — Kgb
- tan
  - sneakers — Kts
  - boots — Ktb
- denim
  - sneakers — Kds
  - boots — Kdb

With the help of the tree diagram, you can see that their are twelve possible outcomes or, in this case, outfits. Assuming that he picks a shirt, a pair of pants, and a pair of shoes at random, the probability that he will pick a blue shirt, denim pants, and sneakers is $\frac{1}{12}$. Another method you can use to find the total number of outcomes is the **Counting Principle.** This method uses multiplication.

number of shirts		number of pants		number of shoes		total number of options
2	×	3	×	2	=	12

Notice you get the same answer. Why then would you go to the trouble of making a tree diagram? Some compound events are more complicated than others. For one this straightforward, the Counting Principle works just fine. However, for a more complicated series of events, you would be better off using a visual tool like a tree diagram. See how you do on the next sample question.

► The menu at *Burger Heaven* has 4 different kinds of burgers: a small hamburger, a small cheeseburger, a double hamburger, and a double cheeseburger. It also has 3 different sizes of French fries: small, medium, and large. Finally, it has 5 different kinds of drinks: cola, root beer, orange, shake, or milk. You select at random a meal that consists of one kind of burger, one size of French fries, and one kind of drink. What is the probability that you will select a double hamburger, a large fries, and a cola?

A $\frac{1}{4}$

B $\frac{1}{12}$

C $\frac{1}{30}$

D $\frac{1}{60}$

Use the Counting Principle to figure out the total number of possible outcomes. You should come up with $4 \times 3 \times 5 = 60$. Only one of these corresponds to the meal selected in the question. Therefore, $P = \frac{1}{60}$, and answer choice (D) is correct.

Combinations and Permutations

In the last question, you used the Counting Principle to figure out the total number of possible outcomes. Each outcome was a different combination of one kind of burger, one size of French fries, and one kind of drink. It did not matter whether you selected the kind of burger first, second, or last.

double hamburger—large fries—cola

large fries—double hamburger—cola

large fries—cola—double hamburger

You still get the same combination. They are merely arranged in a different order. On the other hand, suppose the order of the items in the list **was** important. Then, you would be talking about a permutation. A **permutation** is an arrangement or listing of objects in which order **is** important.

For example, imagine that five runners are entered in a race. Medals are awarded to the first three finishers. How many different ways can the five runners finish first, second, and third? To find the answer, you need to use the permutation formula.

$$P(n, r) = \frac{n!}{(n - r)!}$$

The formula means the number of permutations of n things taken r at a time. What's that exclamation point for? The expression 5! means $5 \cdot 4 \cdot 3 \cdot 2 \cdot 1$. It is read as "five factorial." The expression 3!, or "three factorial," means $3 \cdot 2 \cdot 1$. In the case of the five runners, you need to find $P(5, 3)$.

$$P(5, 3) = \frac{5!}{(5 - 3)!} = \frac{5!}{2!} = \frac{5 \cdot 4 \cdot 3 \cdot 2 \cdot 1}{2 \cdot 1}$$
$$P(5, 3) = \frac{120}{2} = 60$$

There are sixty different permutations, or different ways that five runners can come in first, second, and third.

123	124	125	132	134	135	142	143	145	152	153	154

The above list shows twelve different arrangements in which runner #1 comes in first. The same could be done for runner #2, #3, #4, and #5. Therefore, there are $12 \cdot 5$, or 60, different arrangements.

A **combination** is an arrangement or listing of objects in which order is **not** important. For example, suppose a teacher wishes to break up her class of twenty-seven students into groups of three. How many different groups of three are possible? It does not matter in what order the groups are arranged. For example, a group that consists of students 5, 18, and 27 would still be the same group if you wrote it as 27—18—5 or 18—27—5. The order of the students is not important. To find the answer, you need to use the combination formula.

$$C(n, r) = \frac{P(n, r)}{r!} = \frac{n!}{(n - r)! r!}$$

The formula means the number of combinations of n things taken r at a time. Notice the formula is the number of permutations divided by $r!$. Dividing by $r!$ eliminates the permutations that are the same combination, only in a different order. In this case, you need to find $C(27, 3)$.

Know It All! High School Math

$$C(27, 3) = \frac{P(27, 3)}{3!} = \frac{27 \cdot 26 \cdot 25}{3 \cdot 2 \cdot 1} = \frac{17{,}550}{6} = 2{,}925$$

There are 2,925 different combinations, or 2,925 different groups of three possible in a class of twenty-seven students! That problem was rather complicated. Try the sample question below to check your understanding of combinations.

▶ Mathletes is a school club that competes against other schools in math competitions. The club has 6 seniors. The coach needs to pick 3 seniors to go to the playoffs. How many different groups of 3 seniors are possible?

A   20
B   60
C   120
D   720

You need to find how many different combinations of three seniors can be made from a group of six seniors. In other words, you need to find $C(6,3)$.

$$C(6,3) = \frac{P(6, 3)}{3!} = \frac{6 \cdot 5 \cdot 4}{3 \cdot 2 \cdot 1} = \frac{120}{6} = 20$$

There are twenty different groups of three possible. Therefore, answer choice (A) is correct.

Directions: Read the passage below and answer the question that follows.

Losing the Lotto

The lotto in California works like this: You buy one ticket and it is good for one drawing. On the ticket, you pick six numbers from 1 to 49. You win the jackpot if you correctly pick all six winning numbers. The odds of this happening are 1 in 13,983,816. According to statistician Mike Orkin, this means that if you bought fifty tickets a week, every week, you could expect to win the jackpot once in 5,000 years! If the same amount of money that was spent on lotto tickets were spent on gasoline, it would buy enough gasoline for an economy car to make 730 round trips to the moon and back. Orkin says, "Suppose you're in a football stadium filled with 70,000 people. Then, suppose there are 200 such stadiums. Select one person at random from these 200 stadiums. Your chance of being selected is the same as your chance of winning the lotto jackpot."

▶ The "Daily Number" is a daily lottery game conducted by many states of the union. The winning number is determined by randomly drawing a numbered ball from each of 3 containers. The balls in each container are numbered 0 to 9. There are 1,000 possible winning numbers (10 • 10 • 10). A person picking the winning number wins $600 for a $1 ticket. You decide to buy 10 tickets. On each ticket, you select 3 identical digits, that is, 000, 111, 222, . . . , 999. What is the probability that one of your tickets will have the winning number?

A $\frac{1}{100}$

B $\frac{1}{200}$

C $\frac{1}{500}$

D $\frac{1}{1,000}$

Know It All Approach

Use the **Know It All Approach.** Start by reading the question carefully and making note of the words and information that you need in order to find the correct answer. To determine probability, you need to know the number of ways an event can occur and the number of possible outcomes. The question gives you the number of possible outcomes: 1,000. The event in this case is one of your tickets having the winning number. Since you bought ten tickets, there are ten ways the event can occur.

Next, answer the question. Set up the probability.

$$P = \frac{\text{number of ways that an event can occur}}{\text{number of possible outcomes}} = \frac{10}{1,000} = \frac{1}{100}$$

Now that you have an answer, check your work. A good way to check your answer is to figure out the probability that you will **not** have the winning number. Since you selected ten possible winning numbers out of 1,000, the probability that the winning number will be one of the remaining 990 numbers is $\frac{990}{1,000}$ or $\frac{99}{100}$. The sum of the two probabilities should equal 1. $\frac{1}{100} + \frac{99}{100} = \frac{100}{100} = 1$.

Finally, read all the answer choices, and try to eliminate those that you know are incorrect. You cannot use Process of Elimination here. You can only look at all the answer choices and then hope that one of them agrees with your answer. Answer choice (A) is correct.

One Smart Bird

"How many?" "What color bigger?" "What color smaller?" These are the kinds of questions scientist Irene Pepperberg is asking. The amazing thing is *who* she expects to answer! For the past twenty-two years, she has been teaching Alex, an African grey parrot, a basic form of English. So far, he can recognize fifty different objects by name. He can count up to six, recognize seven different colors, and distinguish among five different shapes. He also understands the concepts of "bigger," "smaller," "same," and "different." In her lab at the University of Arizona, Dr. Pepperberg teaches Alex by showing him objects made of different materials, such as plastic, wood, or felt, and then asking him questions. She says Alex responds correctly about 80 percent of the time, much higher than would be possible by chance alone. Her work deals with one of the most controversial topics in biology: whether animals are capable of abstract thought.

Many scientists believe that animals do not have actual thought processes. Others claim that in many instances, animals appear to be thinking, and therefore further studies should be done. Dr. Pepperberg hopes that her work may someday help children with brain-related illnesses learn how to use language.

Alex the brainy
African Grey Parrot

1. A scientist places a tray holding 7 shapes in front of a parrot. Among the shapes are 2 cubes, 2 spheres, 1 rectangular block, 1 cylinder, and 1 pyramid. The first shape the parrot touches with its beak is a cube. The scientist removes the cube and then waits for the parrot to touch another shape at random. What is the probability that the second shape the parrot touches is also a cube?

   A $\frac{1}{21}$

   B $\frac{1}{7}$

   C $\frac{1}{6}$

   D $\frac{2}{7}$

2. Stacey has the following coins in her purse: 2 quarters, 3 dimes, 1 nickel, and 3 pennies. She reaches into her purse and picks 1 coin without looking or feeling for the size. What is the probability that Stacey will pick a dime?

   A $\frac{1}{9}$

   B $\frac{1}{3}$

   C $\frac{3}{8}$

   D $\frac{3}{7}$

3. Each of the choices listed below describes two events. For which choice are the two events dependent events?

   A You roll a number cube; you roll it again.
   B You pick a card from a deck; you put the card back and then pick another card.
   C You flip a coin; you flip it again.
   D You pick a card from a deck; without putting the card back, you pick another card.

Directions: Read the passage below and answer the question that follows.

Dognosis

Doctors at Cambridge University Veterinary School in England hope to obtain financial backing for a project they call "dognosis." They hope to train dogs to detect prostate cancer by smelling urine samples. This is not the first instance of dogs helping to diagnose cancer. In 1989, a British medical journal recounted the story of a Border collie–Doberman mix that constantly sniffed at a mole on its owner's thigh. The woman had it examined and found out it was a malignant tumor. In another case, a Labrador kept, sniffing at a lesion through his owner's pants. The lesion was a form of skin cancer, and it was removed.

A dermatologist at a clinic in Tallahassee, Florida, began working to train dogs to find samples of melanoma that had already been removed. One dog was 100 percent successful in being able to detect melanoma samples.

4. According to the table below, what is the probability that a woman will develop breast cancer by the age of forty?

A Woman's Chances of Developing Breast Cancer Increases with Age	
By age 30	1 out of 2,212
By age 40	1 out of 235
By age 50	1 out of 54
By age 60	1 out of 23
By age 70	1 out of 14
By age 80	1 out of 10
Ever	1 out of 8

A   0.071
B   0.043
C   0.019
D   0.004

In Chapter 19, you learned that probability is the measure of how likely it is that an event will occur. You discovered how to find the probability of simple and compound events. You also learned to distinguish between independent and dependent events. You found out how to construct a tree diagram and how to use the Counting Principle. Finally, you learned how to calculate the number of possible permutations and combinations in a given situation.

You also know more about the weird and wonderful world waiting for you just beyond your front door.

What are the odds of winning the lotto?
Depending on the type of lottery, the odds can vary. But usually they're pretty bad! In a California state lottery, you need to pick six numbers between 1 and 49. The odds are an astounding one in 13,983,816!

Which bird can count up to six, recognize seven different colors, and distinguish among five different shapes?
Alex the Great, an African Grey Parrot, has shown the intellectual ability that some scientists thought birds didn't have.

In the future, how might some forms of cancer be detected?
Incredibly, sniffing dogs may someday be able to detect skin cancer. Their sense of smell is so strong that they can smell cancerous tumors or melanoma!

Directions: Read the passage below and answer the questions that follow.

Taco Grande

The world's largest taco was created in Houston, Texas, on May 5, 2000. It measured 16 feet long by $2\frac{1}{2}$ feet wide and weighed 1,144 pounds. LaRanchera Inc. made the flour tortilla (the outside covering), while Mama Ninfa's Original Mexican Restaurant blended the taco filler. Inside were 943 pounds of fajita meat, 73.5 pounds of tomatoes, 47 pounds of chopped cilantro, and 81 pounds of chopped onions.

Recently, however, a group of restaurant owners from Mexico claims to have broken that record. The Mexican restaurateurs say their taco was 32 feet long and was made from 480 wheat tortillas stuck together and filled with meat and cheese. One of the restaurant owners said, "The taco is a symbol of Mexico. It's not possible that the United States holds the record."

1. The taco made in Texas was 16 feet long. About how long was the taco in meters?

   A 1.25 m

   B 5 m

   C 48 m

   D 50 m

2. If you flip 3 coins, what is the probability that all 3 coins will land on heads?

   A $\frac{1}{2}$

   B $\frac{1}{2^2}$

   C $\frac{1}{2^3}$

   D $\frac{3}{2^3}$

3. The graph below shows the percentages of different types of movies rented last week at a local video store.

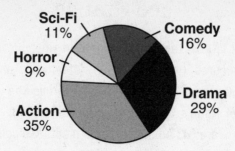

If the video store rented 840 movies last week, how many of them were action movies?

A 35
B 244
C 294
D 805

4. The numbers listed below are test scores for a class of twenty students.

   93, 84, 97, 81, 78, 86, 85, 92, 72, 81, 55, 91, 84, 90, 75, 94, 83, 60, 81, 95

   Part A Find the mean.

   Part B Find the median.

   Part C Find the mode.

---

5. The table below shows how much each of four teenagers spent for dinner at a restaurant. What is the mean amount spent by each teenager?

	Alexis	Carlos	Tracie	Diana
appetizer	$4.50	$3.95		$5.45
main course	$13.95	$11.50	$10.75	$12.50
dessert	$4.00	$5.25	$4.95	
drink	$2.15	$3.00	$1.95	$2.50
Total	$24.60	$23.70	$17.65	$20.45

A $19.82
B $21.50
C $21.60
D $22.08

Directions: Read the passage below and answer the questions that follow.

Monumental Memorial

What is most likely the world's largest finger painting was crafted on September 10, 2002, in Southfield, Michigan. Forty-three students from the Lawrence Technological University let their fingers do the painting, with no help from brushes, rollers, or sponges. The painting of a giant American flag had an area of 2,766.5 square feet (257 square meters). It was produced in memory of those lost in the terrorist attacks on September 11, 2001.

6. If each of the forty-three students painted the same area of the giant American flag, about how many square meters did each student paint?

   A $6 \, \text{m}^2$
   B $25.7 \, \text{m}^2$
   C $46 \, \text{m}^2$
   D $64 \, \text{m}^2$

7. Alicia rolls two cubes whose sides are numbered 1 through 6. What is the probability that she will roll a double (that is, 1–1, 2–2, 3–3, 4–4, 5–5, or 6–6)?

   A $\dfrac{1}{12}$
   B $\dfrac{1}{6}$
   C $\dfrac{1}{4}$
   D $\dfrac{1}{3}$

Answer Key for Chapter Questions

Chapter 2
1. C
2. D
3. C
4. D
5. C
6. 200 million
7. 3,500

Chapter 3
1. A
2. 13.5
3. D
4. 480 ft.
5. B

Chapter 4
1. C
2. D
3. A
4. B
5. 96%

Chapter 5
1. C
2. 4,000 mi.
3. 48 m
4. B
5. C

Chapter 6

1. B
2. C
3. D
4. B
5. D
6. B
7. A

Brain Booster #1

1. C
2. B
3. C
4. A
5. C
6. D
7. $40
8. C
9. D
10. D
11. A
12. C
13. C
14. A
15. C

Chapter 7

1. 37 seconds
2. A
3. D
4. C
5. A
6. B
7. C
8. D
9. D
10. A

Chapter 8

1. C
2. Table A
3. D

Chapter 9

1. B
2. D
3. C
4. C
5. D
6. $12
7. B

Chapter 10

1. $x = 2, y = 6$

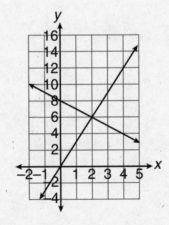

2. C

3.   400 ft.

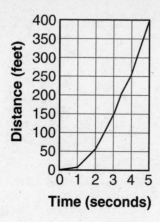

Chapter 11

1. $5x - 1 = 4x + 2$

   $x = 3$

   $3x = 9$

2. C
3. B
4. $4x - 2 = 2x + 8$

   $2x = 10$

   $x = 5$

   Side 1: $4(5) - 2 = 18$

   Side 2: $5 + 13 = 18$

   Side 3: $2(5) + 8 = 18$

   Side 4: $3(5) + 3 = 18$

   The figure is a rhombus
   because all four sides are equal.
5. C
6. $144°$

Brain Booster #2

1. C
2. $3.29
3. A
4. C
5. B
6. D
7.

x	$y = x^2 + 6x + 8$	y
-2	$y = -2^2 + 6(-2) + 8$	0
-1	$y = -1^2 + 6(-1) + 8$	3
0	$y = 0^2 + 6(0) + 8$	8
1	$y = 1^2 + 6(1) + 8$	15
2	$y = 2 + 6(2) + 8$	24

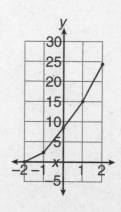

Chapter 12

1. B
2. C
3. B
4. $C = 2\pi r$
   $C = 2\,(3.14)(25)$
   $C = 157$ m
   $P = 90 + 90 + 157$
   $P = 337$ m
   $d = 4.75P$
   $d = 4.75\,(337)$
   $d = 1{,}600.75$ m
5. C
6. A
7. C

Chapter 13

1. 1,350 kg
2. C
3. C
4. C
5. $V = 13{,}200 \div 264$
   $V = 50$ m^3
   $V = \pi r^2 h$
   $50 = 3.14\,r^2(1.3)$
   $50 = 4.082 r^2$
   $12.25 = r^2$
   3.5 m $= r$

Chapter 14

1. C
2. C
3. A
4. 27 ft.

Chapter 15

1. C
2. C
3. D
4. $160 - 110 = 50$
   $c^2 = a^2 + b^2$
   $c^2 = 120^2 + 50^2$
   $c^2 = 14,400 + 2,500$
   $c^2 = 16,900$
   $c = 130$ ft.

Brain Booster #3

1. Part A SA = 75.36 cm^2
   Part B V = 28.36 cm^3
2. B
3. 40 m
4. D
5. C

Chapter 16

1. D
2. C
3. B
4. D
5. A
6. B
7. C

Chapter 17

1. D
2. C
3. B
4. B
5. D
6. C

Chapter 18

1. A
2. D
3. positive correlation

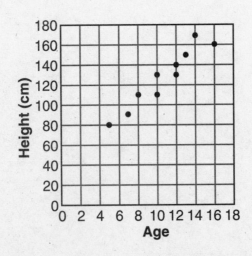

4. A
5. B

Chapter 19

1. C
2. B
3. D
4. D

Brain Booster #4

1. B
2. C
3. C
4. mean = 82.85
   median = 84
   mode = 81
5. C
6. A
7. B

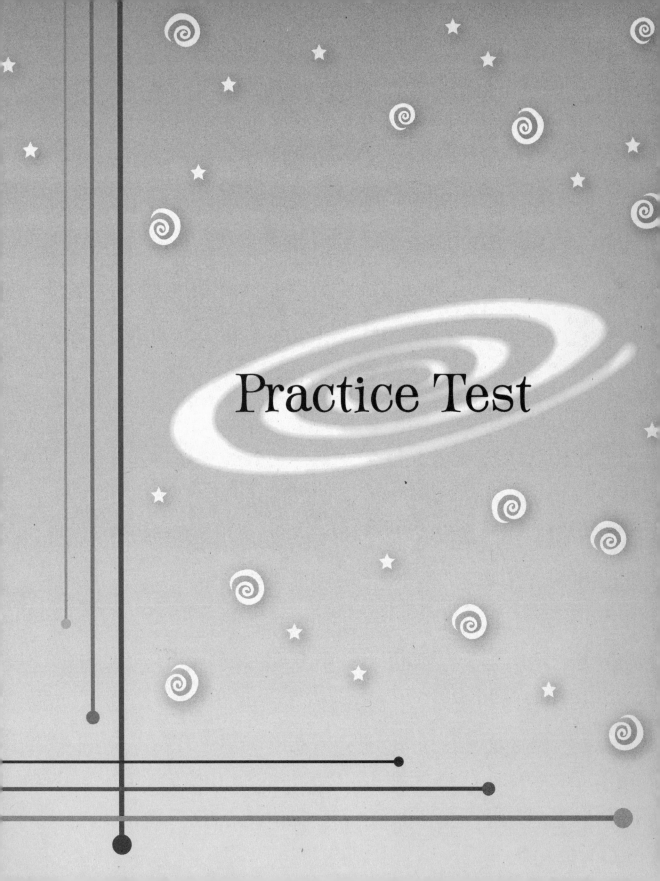

Practice Test

Introduction to the Practice Test

By now you've reviewed all the important skills you should know for high school math. You know how to compute with fractions, decimals, and percents (Chapter 4). You know the difference between solving equations and solving inequalities (Chapter 9). You also know how to find the area, perimeter, surface area, and volume of various figures (Chapters 12 and 13). And these are just a few examples that don't even include all the excellent tidbits of information you've picked up. You *Know It All!*

If you're ready, it's time to try out the skills from the nineteen chapters in this book in a practice test. This test may be similar to a test you take in class. It contains multiple-choice, short-answer, and open-response questions.

Each multiple-choice question has four answer choices. You should fill in the bubble for the correct answer choice on the separate answer sheet. Cut or tear out the answer sheet on page 291, and use it for the multiple-choice questions. You can write your answers to the short-answer and open-response questions directly onto the test.

The practice test contains thirty-six questions. Give yourself ninety minutes to complete the test.

The *Know It All!* Math Reference Sheet on the following page includes mathematical formulas that you may use to answer questions on the practice test. Use the reference sheet for questions that require formulas to answer them.

Take the practice test the same way you would take a real test. Don't watch television, don't talk on the telephone, and don't listen to music while you take the test. Sit at a desk with a few pencils, and have an adult time you, if possible. Take the test in one day and all in one sitting. If you break up the test in parts, you won't get a real test-taking experience.

When you've completed the practice test, you may go to page 283 to check your answers. Each question also has an explanation to help you understand how to answer it correctly. Don't look at this part of the book until you've finished the practice test.

Good luck!

Know It All! Math Reference Sheet

The *Know It All!* Math Reference Sheet below is similar to one you may see on a standardized-achievement test. Use the reference sheet to answer some of the questions on the practice test.

Abbreviations

b = base of a polygon

h = height

l = length

w = width

A = area

C = circumference

r = radius

d = diameter

S = surface area

V = volume

m = slope

For π, use 3.14 or $\frac{22}{7}$.

Formulas

Triangle: $\mathbf{A} = \frac{1}{2}\mathbf{bh}$

Parallelogram: $\mathbf{A} = \mathbf{bh}$

Rectangle: $\mathbf{A} = \mathbf{lw}$

Circle: $\mathbf{C} = 2\pi\mathbf{r}$

$\mathbf{A} = \pi\mathbf{r}_2$

Rectangular solid:
$\mathbf{S} = 2(\mathbf{wh} + \mathbf{lh} + \mathbf{lw})$

$\mathbf{V} = \mathbf{lwh}$

Cylinder: $\mathbf{S} = 2\pi\mathbf{rh} + 2\pi\mathbf{r}_2$

$\mathbf{V} = \pi\mathbf{r}_2\mathbf{h}$

Slope Formula: $\mathbf{m} = \dfrac{\mathbf{y}_2 - \mathbf{y}_1}{\mathbf{x}_2 - \mathbf{x}_1}$

Pythagorean theorem:
$\mathbf{c}_2 = \mathbf{a}_2 + \mathbf{b}_2$

Distance = rate • time

Interest = principal • rate • time

Practice Test

1. Add the following: $\frac{1}{3} + \frac{1}{5}$

    A $\frac{1}{15}$

    B $\frac{1}{8}$

    C $\frac{6}{15}$

    D $\frac{8}{15}$

Joker

2. It may seem strange, but there have been reported cases of storms raining down animals such as fish and toads. What if someone counted the frogs that fell on her front porch during such a storm?

Minutes of the Storm	Number of Frogs
1	4
2	8
3	16
4	32

If the pattern of the number of frogs falling from the sky during the storm continues, how many frogs will fall during the fifth minute?

A 32

B 34

C 64

D 80

3. Cheetahs are the fastest land animals on earth, capable or reaching top speeds of more than 72 kilometers per hour. Which of the following is the same rate of speed?

A   20 meters per second
B   120 meters per minute
C   12 kilometers per minute
D   200 meters per second

4. From July 1969 through December 1972, NASA's Apollo space program sent men to the moon. Simplify the following absolute value problem to determine the number of men that NASA has successfully sent to the moon.

$$6 + |5 - 2| + |-4| - |-1|$$

A   4
B   5
C   12
D   13

5. Glenn's father wants to build a deck around one-half of his circular backyard pool, and he drew a design of the pool and the deck. Before he buys the wood to build the deck, he needs to know the area of the deck. What is the area of the deck represented by the shaded region in the diagram below? **Show all work.**

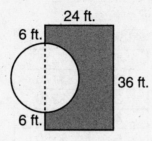

6. Raymond needed to save $200 each month for six months to pay for a motorized skateboard that he wanted. The skateboard, called Exkate Powerboard, is battery-powered and can go up to 22 miles per hour in only 4 seconds. After five months, he had put away the amounts shown in the table below.

Month	Amount
1	$215
2	$175
3	$240
4	$205
5	$180
6	?

How much does Raymond need to put away in the sixth month to reach the required total?

A $185
B $195
C $215
D $225

7. In early 2003, scientists discovered that the speed of gravity is equal to the speed of light. Albert Einstein stated this fact in his 1915 general theory of relativity. The speed of light in a vacuum is 299,800,000 meters per second. How should this number be expressed in scientific notation?

A 2.998×10^{-8}
B 2.998×10^{8}
C 29.98×10^{7}
D 299.8×10^{6}

8. Katrina received $100 in graduation gifts. She invested the money in two savings accounts. After one year, one savings account earned 8% interest and the other savings account earned 12% interest. In total, the two savings accounts gathered $9 in interest for Katrina. This is represented by the system of equations shown below, in which a represents the amount invested in the savings account earning 8% interest and b represents the amount invested in the savings account earning 12% interest.

$a + b = \$100$

$0.08a + 0.12b = \$9$

Solve the system of equations to determine a, the amount Katrina put into the savings account that earned 8% interest.

_____

9. The oldest ballpark in the country is Fenway Park in Boston, Massachusetts. It is also the smallest park, holding about 33,000 people. On weekends, LaTrell works at the concession stand at the baseball stadium. At last Sunday's game, imagine that the ratio of Red Sox baseball caps sold at a concession stand compared to baseball caps sold for all other teams sold was 8 to 3. If a total of 55 baseball caps were sold, how many were for the Red Sox?

A 15
B 20
C 32
D 40

10. When milk goes bad, you usually throw it out. But not cheese makers! When milk goes bad, it produces something called curd, which is then used to create cheese. So cheese is actually made from spoiled milk!

The graph below shows the number of pounds of different kinds of cheeses sold at a grocery store last week.

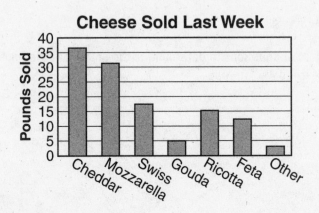

About how much more mozzarella cheese was sold last week than ricotta cheese?

A 12 pounds

B 16 pounds

C 21 pounds

D 31 pounds

11. To get to the top of a ride at an amusement park, people walk up stairs to a platform 20 feet above the ground and then board a cable car. The top of the ride is 170 feet high. The bottom of the ride is 200 feet from the base of the platform. How long is the cable (*x*)? **Show every step of your procedure to determine the answer.**

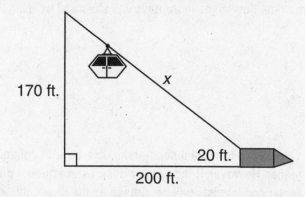

170 ft.

x

20 ft.

200 ft.

12. No one is certain how many different types of organisms exist on Earth. Scientists, however, have described 1.75 million of them and estimate there are millions more. Marisol has to look up the scientific name of 120 different organisms. She spends $\frac{1}{2}$ hour every day working on the project. It took her 3 days to look up 15 organisms. If she continues to work at the same rate, how many more days will she need to finish the project?

    A 21
    B 24
    C 30
    D 36

13. Of five finalists selected to receive a scholarship to a certain college, only one is chosen as the winner. However, if the winner decides to attend a different college, one of the four remaining finalists will be chosen as the new winner. The diagram below illustrates the probabilities of winning and losing.

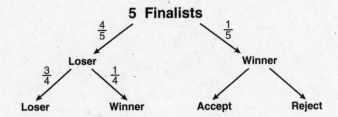

What is the probability that a student will lose the first round and win the second round?

14. In the figure below, line *a* is parallel to line *b*. What is the value of *x*?

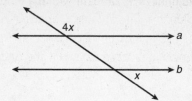

A 20°
B 30°
C 36°
D 45°

15. Mrs. Garcia leased a car for $299 a month plus $0.15 a mile for every mile that exceeds 1,000 miles per month. To figure out her total cost, she uses the expression $C = 299n + 0.15(m - 1,000n)$, where C = total cost for n months, n = the number of months leased, and m = the total number of miles driven. If she leased the car for 24 months and drove it 27,500 miles, what was her total cost?

16. Darren's parents bought a sandbox for the backyard. It measures 4 meters long, 2 meters wide, and 1 meter high. They need to buy sandbags to fill it with sand. If each sandbag contains 2 cubic meters of sand, how many sandbags do they need to buy?

A   2
B   4
C   8
D   16

17. Construct a graph of the equation $y = 2x - 3$. Fill in the table of values below for $x = -1$ to $x = 4$, and then plot the points on the coordinate grid provided. Connect the points with a straight line.

x	y = 2x − 3	y
−1		
0		
1		
2		
3		
4		

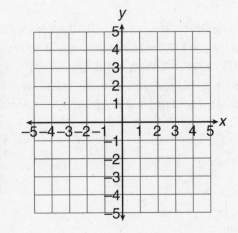

Know It All! High School Math

18. The black widow spider is a small poisonous, black spider, usually no more than an inch and a half long. It's called the black widow spider because the female spider eats her mate after mating.

You can usually spot a female black widow spider by the bright red hourglass shape on the underside of her abdomen. In the figure below, $\overline{AE}$ and $\overline{BD}$ bisect each other at point C. Of the following, which theorem could you use to prove that $\triangle ABC \cong \triangle DEC$?

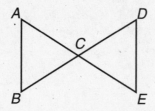

A AAS
B ASA
C SAS
D SSS

19. In the figure below, △BEC is inscribed in square ABCD. If the perimeter of square ABCD is 64 inches, what is the area of △BEC?

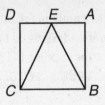

A   32 sq. in.
B   64 sq. in.
C   128 sq. in.
D   256 sq. in.

20. Solve for x.

$5x - 3 = 2x + 18$

A   −7
B   −5
C   5
D   7

21. Robert Wadlow, of Alton, Illinois, is often referred to by the title "tallest man in history." By the time he was twenty years old, he was all of 8 feet, 11.1 inches tall.

The table below lists the frequency of the different heights of a class of thirty students. What is the mode of this data set?

Height	Frequency
4 ft. 11 in.	1
5 ft. 0 in.	0
5 ft. 1 in.	3
5 ft. 2 in.	4
5 ft. 3 in.	7
5 ft. 4 in.	3
5 ft. 5 in.	4
5 ft. 6 in.	3
5 ft. 7 in.	2
5 ft. 8 in.	2
5 ft. 9 in.	1

A   5 ft. 2 in.
B   5 ft. 3 in.
C   5 ft. 4 in.
D   5 ft. 5 in.

22. Which of the following tables represents the function $f(x) = -2x + 5$?

x	f(x)
−1	3
0	5
1	7
2	9

A

x	f(x)
−1	−7
0	−5
1	−3
2	−1

B

x	f(x)
−1	−3
0	−1
1	1
2	3

C

x	f(x)
−1	7
0	5
1	3
2	1

D

23. The world record for the long jump was broken in 1988 with a 24.67-foot leap. At a recent track meet, the winning distance in the long jump was 23 feet. About how far is this distance in meters?

A 7
B 7.5
C 8
D 8.5

24. In 1933, the U.S. Mint made a $20 gold coin called the Double Eagle. Because of the Great Depression, however, President Franklin Roosevelt ordered the coins be melted down to create a stockpile of gold. Only one Double Eagle coin exists today! It was sold in 2001 for $7.59 million.

A student flips three coins at the same time. Draw a tree diagram to illustrate all the possible combinations. Then, determine the probability of getting two heads and one tail.

25. The circle graph below shows Mr. and Mrs. Patterson's monthly expense budget by categories.

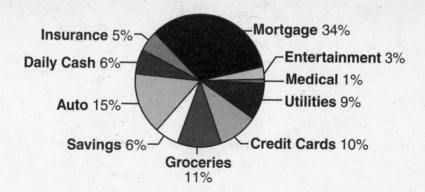

If their total expenses equal $3,600, how much do they spend each month on groceries?

A $360
B $396
C $540
D $720

26. Jason paid $87.25 for five black-and-white ink cartridges and $57.00 for two color ink cartridges. How much more did he pay for each color ink cartridge than each black-and-white ink cartridge?

A $5.55
B $7.25
C $10.75
D $11.05

27. Which of the following solids is **NOT** a prism?

A   B   C   D

28. Given the formula $F = ma$, if $F = 160$ and $m = 5$, what is the value of a?

   A 16
   B 32
   C 40
   D 80

29. At the end of last summer, the lilac bush in Lenora's backyard was 39 inches tall. Since that time, it has grown 13 inches. How tall is the lilac bush now?

   A   3 ft. 11 in.
   B   4 ft. 2 in.
   C   4 ft. 4 in.
   D   4 ft. 6 in.

30. In 1998, a certain brand of sport utility vehicle got an average of 14 miles per gallon of gasoline. In 2003, the same model got an average of 17.5 miles per gallon. What is the percent increase in gas mileage?

31. The hourly temperature change over a 12-hour period is shown on the graph below.

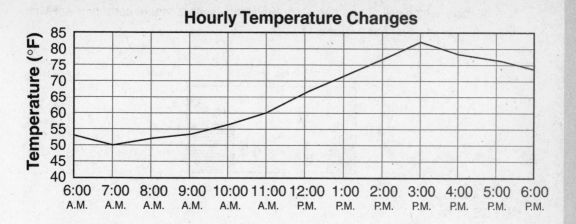

During which 3-hour period did the temperature change the most?

A 6:00 A.M.–9:00 A.M.

B 10:00 A.M.–12:00 P.M.

C 12:00 P.M.–3:00 P.M.

D 3:00 P.M.–6:00 P.M.

32. Mrs. Chaudri wants to put heavy-duty carpeting in her basement. The floor plan is shown in the diagram below. The shaded areas represent the stairway and the oil burner room, which will not be carpeted. How many square feet of carpeting will Mrs. Chaudri need to cover her basement floor?

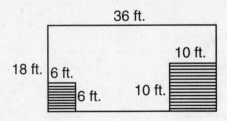

A   512 sq. ft.
B   536 sq. ft.
C   584 sq. ft.
D   648 sq. ft.

33. During football games, the coach provides the players with a cylindrical cooler filled with a sports drink. The cooler has a radius of 20 centimeters and a height of 100 centimeters. A science handbook states that 3,790 cubic centimeters is equal to one gallon. About how many gallons does the cooler hold?

A   8
B   15
C   24
D   33

34. The graph of a linear equation is shown in the figure below. What is the slope of the line?

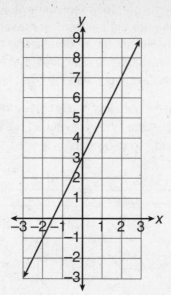

A   −2

B $-\dfrac{1}{2}$

C $\dfrac{1}{2}$

D   2

35. Simplify: $3 \bullet 5^2 \div 15\,(7 - 2^3)$

A   −5

B   4

C   6

D   12

Joker

36. Sarah L. Winchester, eccentric heiress to the Winchester fortune, spent her entire life building what is now known as the Winchester Mystery House. This four-story, 160-room mansion is full of secret doors, dead-end stairways, and other curiosities. The reason for all these features is as odd as the house itself. Sarah Winchester believed she was living under a curse, and, attempting to foil spirits that might come looking for her, she built a house that even ghosts would get lost in.

Look at the blueprint below. The actual width, left to right, of the house is 42 feet. Based on the scale drawing of the house shown below, what is the actual depth, top to bottom?

A   34 ft.

B   36 ft.

C   38 ft.

D   40 ft.

Answers and Explanations to the Practice Test

1. **D** To add fractions, you need to change them to equivalent fractions that have the same denominator. The lowest common denominator for $\frac{1}{3}$ and $\frac{1}{5}$ is 15. You need to change each fraction to an equivalent fraction with a denominator of 15. $\frac{1}{3}$ becomes $\frac{5}{15}$. $\frac{1}{5}$ becomes $\frac{3}{15}$. $\frac{5}{15} + \frac{3}{15} = \frac{8}{15}$.

2. **C** Follow the pattern. By the fifth minute, 64 frogs will have fallen.

3. **A** The first step is to convert 72 kilometers per hour to meters per hour. Because 1 kilometer = 1,000 meters, this means that 72 kilometers per hour = 72,000 meters per hour. The next step is to convert 72,000 meters per hour to meters per minute. 1 hour = 60 minutes, so you must divide by 60. Thus, 72,000 ÷ 60 = 1,200 meters per minute. The final step is to convert meters per minute to meters per second. Because 1 minute = 60 seconds, you must divide by 60 again. 1,200 ÷ 60 = 20 meters per second.

4. **C** The absolute value of a number is always positive. The expression becomes $6 + 3 + 4 - 1 = 12$.

5. **Area of deck = 637.92 or 638 sq. ft.** The area of the rectangle is $A = lw = 24 \cdot 36 = 864$ square feet. However, the right half of the pool cuts into the rectangle. Therefore, to find the area of the deck, the area of the rectangle must be reduced by the area of half the pool. The diameter of the pool is 36 feet minus the two 6-foot sections, or 24 feet. The radius is one-half the diameter, or 12 feet. The area of the entire pool (circle) is $A = \pi r^2 = 3.14 \cdot 12^2 = 3.14 \cdot 144 = 452.16$ square feet. The area of **half** the pool is $452.16 \cdot \frac{1}{2} = 226.08$ square feet. Therefore, the area of the deck equals $864 - 226.08$, or 637.92.

6. **A** The total amount of money Raymond needed at the end of six months was $200 \cdot 6 = $1,200. According to the table, he has $215 + $175 + $240 + $205 + $180 = $1,015. Therefore, in the sixth month, he needs to put away $1,200 − $1,015 = $185.

7. **B** To write 299,800,000 in scientific notation, you must move the decimal point until it is between the 2 and the first 9. Thus, you must move the decimal point 8 places to the left. Large numbers have positive exponents, while small numbers have negative exponents. You can eliminate answer choice (A). Therefore, the answer is 2.998×10^8. Although answer choices (C) and (D) are mathematically correct, they do not conform to the accepted practice of moving the decimal point until you end up with a number between 1 and 10.

8. **$a = \$75$** Use the substitution method. If $a + b = 100$, then $b = 100 - a$.

$0.08a + 0.12b = 900$

$0.08a + 0.12(100 - a) = 99$

$0.08a + 12 - 0.12a = 9$

$-0.04a + 12 = 9$

$-0.04a = -3$

$a = 75$

9. **D** You need to find two numbers that add up to 55 and are in a ratio of 8 to 3. If you add 8 and 3, you get 11. Divide 55 by 11 to get 5. Next, multiply 5 times 8, and then 5 times 3. You should get 40 and 15. These two numbers add up to 55 and are in an 8 to 3 ratio. Therefore, the number of Red Sox baseball caps sold was 40.

10. **B** According to the graph, the amount of mozzarella cheese sold last week was 31 pounds. The amount of ricotta cheese sold last week was 15 pounds. Therefore, $31 - 15 = 16$ pounds more mozzarella cheese was sold last week than ricotta cheese.

11. **$x = 250$ ft.** The first step is to find the height of the right triangle. You need to subtract the height of the platform from the height of the ride. Hence, the height of the right triangle equals $170 - 20$, or 150 feet. The base of the right triangle is 200 feet, or the distance between the ride and the platform. The final step is to solve for x using the Pythagorean theorem: $c^2 = a^2 + b^2$.

$x^2 = 150^2 + 200^2$

$x^2 = 22{,}500 + 40{,}000$

$x^2 = 62{,}500$

$x = 250$ ft.

12. **A**   If it took Marisol 3 days to look up 15 organisms, she is working at a rate of 5 organisms per day. That means it will take her $120 \div 5 = 24$ days to look up 120 organisms. This is **not** the answer. Remember, she has already worked for 3 days. The question asks how many more days will she need to finish the project. Therefore, the answer is $24 - 3$, or 21 days.

13. $P = \dfrac{1}{5}$ **or 0.20 or 20%**   The probability of both event A and event B occurring is the product of their individual probabilities. Thus, $P(\text{A and B}) = P(\text{A}) \cdot P(\text{B})$. According to the diagram, the probability of losing the first round is $\dfrac{4}{5}$. The probability of winning the second round is $\dfrac{1}{4}$. Therefore, the probability of both events occurring is $\dfrac{4}{5} \cdot \dfrac{1}{4}$, or $\dfrac{4}{20}$, which reduces to $\dfrac{1}{5}$.

14. **C**   The two angles, x and $4x$, are supplementary. That is, their sum is 180°. Knowing this, you can set up an equation to solve for x.

$x + 4x = 180$

$5x = 180$

$x = 36$

15. $C = \$7{,}701$ This is a substitution problem. The number of months leased, n, is 24. The number of miles driven, m, is 27,500.

$C = 299n + 0.15(m - 1{,}000n)$

$C = 299 \cdot 24 + 0.15(27{,}500 - 1{,}000 \cdot 24)$

$C = 299 \cdot 24 + 0.15(27{,}500 - 24{,}000)$

$C = 299 \cdot 24 + 0.15 \cdot 3{,}500$

$C = 7{,}176 + 525$

$C = 7{,}701$

16. **B**   First, you need to find the volume of the sandbox. $V = lwh = 4 \cdot 2 \cdot 1 = 8$ cubic meters. If each sandbag contains 2 cubic meters of sand, then they need $8 \div 2$, or 4 sandbags.

17. This question has two parts. The first part is to fill in the table with the correct values for *x* and *y*. To do this, substitute each *x*-value (from −1 through 4) into the equation in the middle column of the table. Then, solve the equation to find each *y*-value. Write each answer in the right-hand column. The second part is to plot the points on a graph and connect them with a straight line.

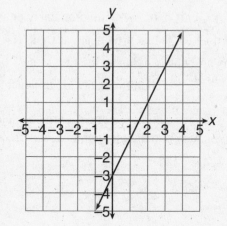

x	y = 2x − 3	y
−1	$y = 2(-1) - 3$	−5
0	$y = 2(0) - 3$	−3
1	$y = 2(1) - 3$	−1
2	$y = 2(2) - 3$	1
3	$y = 2(3) - 3$	3
4	$y = 2(4) - 3$	5

18. **C** If $\overline{AE}$ and $\overline{BD}$ bisect each other at point *C*, then $\overline{AC} \cong \overline{EC}$ and $\overline{BC} \cong \overline{DC}$. You should also realize that $\angle ACB \cong \angle DCE$ because they are vertical angles. Hence, for each triangle, two sides and the included angle are congruent. Therefore, the two triangles are congruent by side-angle-side (SAS).

19. **C** To find the area of $\triangle BEC$, you need to know the lengths of its base and its height. Since $\triangle BEC$ is inscribed in a square, both its base and its height are equal to the length of one side of the square. To find the length of one side of the square, divide the perimeter by 4. Thus, $64 \div 4 = 16$ inches. The area of $\triangle BEC$ is $A = \frac{1}{2} bh = \frac{1}{2} \cdot 16 \cdot 16 = 128$ square inches.

20. **D** The most important thing you need to do is isolate the variable on one side of the equation.

$5x - 3 = 2x + 18$

$3x - 3 = 18$

$3x = 21$

$x = 7$

21. **B** This is an easy question if you know what the mode is. The mode is the number that appears most often in a set of data. According to the frequency table, the height that appears most often (7 times) is 5 feet 3 inches.

22. **D** To find the correct answer, substitute the *x*-values (−1 through 2) in the equation to find the value of $f(x)$. The data in the last column, $f(x)$, agree with answer choice (D).

23. **A** You need to have a general idea of how many feet are in a meter. A meter equals 39.37 inches. 3 feet equal 36 inches. Hence, a meter is about 3 feet 3 inches, or $3\frac{1}{4}$ feet. If you divide 23 feet by $3\frac{1}{4}$, you should get $7\frac{1}{13}$ meters, which you can round off to 7 meters.

24. **P(2 heads and 1 tail)** $= \frac{3}{8}$ **or 0.375 or 37.5%** To show the possible outcomes of flipping three coins at the same time, you will need to draw a tree diagram with three sets of branches.

1st Coin	2nd Coin	3rd Coin	Outcome
		H	HHH
	H	T	HHT
		H	HTH
H	T	T	HTT
		H	THH
	H	T	THT
H		H	TTH
	T	T	TTT

After you have filled in the "Outcome" column in the tree diagram, notice that three of the eight possible outcomes involve getting two heads and one tail. Therefore, *P*(2 heads and 1 tail) $= \frac{3}{8}$ or 0.375 or 37.5%.

25. **B**   According to the circle graph, groceries make up 11% of Mr. and Mrs. Patterson's monthly expenses. If their total monthly expenses equal $3,600, then they spend 11% of $3,600 on groceries. Therefore, they spend 0.11 • 3,600, or $396, each month on groceries.

26. **D**   You need to find the cost of one black-and-white ink cartridge. Then, you need to find the cost of one color ink cartridge. Finally, you need to subtract to find out how much more one color ink cartridge costs than one black-and-white ink cartridge.

    $87.25 ÷ 5 = $17.45 per black-and-white ink cartridge

    $57.00 ÷ 2 = $28.50 per color ink cartridge

    $28.50 − $17.45 = $11.05

27. **C**   A prism is a three-dimensional figure that has two parallel and congruent bases in the shape of a polygon. The other faces of a prism are rectangles. The cone, answer choice (C), does not have two bases, nor does it have rectangular faces.

28. **B**   This is a substitution problem. Given the equation $F = ma$, substitute 160 for F and 5 for m. Hence, $160 = 5a$. Divide both sides by 5, and you get $a = 32$.

29. **C**   This problem has two steps. The first step is to add the growth of 13 inches to the previous height of 39 inches. The plant is $39 + 13$, or 52, inches tall. The second step is to convert this height to feet and inches. Since 1 foot = 12 inches, you can divide 52 by 12. You should get 4 feet with 4 inches left over. Therefore, the lilac bush is now 4 feet 4 inches tall.

30. **25%**   Remember that percent increase or decrease is based on the original amount. In this problem, the original mileage is 14 miles per gallon. The new mileage is 17.5 miles per gallon. The increase in mileage is $17.5 − 14 = 3.5$ miles per gallon. When figuring out the percent increase, divide the difference by the original mileage. Hence, the percent increase in mileage is $3.5 ÷ 14$, or 0.25, which equals 25%.

31. **C**   The data in the graph cover a 12-hour period. The four answer choices are four different 3-hour periods. The best way to answer this question is to figure out the temperature change for each 3-hour period.

A   6:00 A.M.–9:00 A.M. $53 - 52 = 1$-degree change

B   10:00 A.M.–12:00 P.M. $66 - 53 = 13$-degree change

C   12:00 P.M.–3:00 P.M. $82 - 66 = 16$-degree change

D   3:00 P.M.–6:00 P.M. $82 - 74 = 8$-degree change

The 3-hour period during which the temperature changed the most is 12:00 P.M.–3:00 P.M., or answer choice (C).

32. **A**   The area of the entire basement is $A = lw = 18 \cdot 36 = 648$ square feet. However, this is not the answer. You need to subtract the areas of the stairway and the oil burner room, which will not be carpeted. The stairway measures 6 feet. by 6 feet., so its area is 36 square feet. The oil burner room measures 10 feet by 10 feet, so its area is 100 square feet. Therefore, the area that will be carpeted equals $648 - (36 + 100)$, or $648 - 136$, which equals 512 square feet.

33. **D**   This problem has two steps. The first step is to find the volume of the cylindrical cooler.

$V = \pi r^2 h$

$V = 3.14 \cdot 20^2 \cdot 100$

$V = 3.14 \cdot 400 \cdot 100$

$V = 125{,}600 \text{ cm}^3$

The second step is to figure out how many gallons the drum holds. Since 3,790 cubic centimeters equals 1 gallon, the drum holds $125{,}600 \div 3{,}790 = 33.14$ gallons. You can round the answer off to 33 gallons.

34. **D**   The slope of a line is "the rise over the run," or the vertical change divided by the horizontal change. You can start anywhere along the line. However, it is probably easier to start near an axis. Locate the point (0, 3) on the graph, and then follow the line to the end, point (3, 9). The vertical distance went from 3 to 9, for a change of 6 units. The horizontal distance went from 0 to 3, or 3 units. Hence, the slope is 6 ÷ 3, or 2. You could also use the formula for slope. Let (0, 3) represent x_1 and y_1, and let (3, 9) represent x_2 and y_2.

$$m = \frac{y_2 - y_1}{x_2 - x_1}$$

$$m = \frac{9 - 3}{3 - 0}$$

$$m = \frac{6}{3}$$

$$m = 2$$

35. **A**   To simplify this expression, you must use the order of operations.

$3 \cdot 5^2 \div 15(7 - 2^3)$

$3 \cdot 5^2 \div 15(7 - 8)$

$3 \cdot 5^2 \div 15(-1)$

$3 \cdot 25 \div 15(-1)$

$75 \div 15(-1)$

$5(-1)$

-5

36. **C**   Scale diagrams and models are similar figures. Similar figures have the same shape but different sizes. The lengths of the sides of similar figures are proportional. Set up a proportion. Then, cross multiply. This gives you the expression $5\frac{1}{4} x = 42 \cdot 4\frac{3}{4}$. Solve the expression and you get $x = 38$ feet.

Answer Sheet

1. (A) (B) (C) (D)
2. (A) (B) (C) (D)
3. (A) (B) (C) (D)
4. (A) (B) (C) (D)
5. Use space provided.
6. (A) (B) (C) (D)
7. (A) (B) (C) (D)
8. Use line provided.
9. (A) (B) (C) (D)
10. (A) (B) (C) (D)
11. Use space provided.
12. (A) (B) (C) (D)
13. Use line provided.
14. (A) (B) (C) (D)
15. Use line provided.
16. (A) (B) (C) (D)
17. Use space provided.
18. (A) (B) (C) (D)

19. (A) (B) (C) (D)
20. (A) (B) (C) (D)
21. (A) (B) (C) (D)
22. (A) (B) (C) (D)
23. (A) (B) (C) (D)
24. Use space provided.
25. (A) (B) (C) (D)
26. (A) (B) (C) (D)
27. (A) (B) (C) (D)
28. (A) (B) (C) (D)
29. (A) (B) (C) (D)
30. Use line provided.
31. (A) (B) (C) (D)
32. (A) (B) (C) (D)
33. (A) (B) (C) (D)
34. (A) (B) (C) (D)
35. (A) (B) (C) (D)
36. (A) (B) (C) (D)

The Princeton Review

Partnering With You to Measurably Improve Student Achievement

Our proven 3-step approach lets you **assess** student performance, **analyze** the results, and **act** to improve every student's mastery of skills covered by your State Standards.

Assess
Deliver formative and benchmark tests

Analyze
Review in-depth performance reports and implement ongoing professional development

Act
Utilize after school programs, course materials, and enrichment resources